SYNAPSE OF ASH

CYBER HUNTER ORIGINS BOOK 1

D. B. GOODIN

For more information about the Cyber Hunter Origins series visit:

www.cyberhunterorigins.com

www.dbgoodinbooks.com

www.davidgoodinauthor.com

ISBN: 978-1-7350736-2-0 (Paperback)

Some see them as identically birthed twins. I see them as weapons of choice.
 —Dr. Elizabeth Ash

CHAPTER 1

THE CYBORG AWOKE in a small and dark place. From its vantage point, it appeared to be lying on a table. A bright light shone overhead. As the cyborg filled its lungs with the room's cold, stale air, an acrid smell forced the cyborg to turn its head away.

Where am I? the cyborg thought. How did I get here? Who am I?

The cyborg tried sitting up; its arms were weak and started shaking. After a few moments, the cyborg rolled off the table and fell onto the floor. It reflexively tried extending its arms, but a pang of extreme pain shot through its body as its palms struck the concrete floor.

The cyborg started shivering.

Rest some more, it told itself. I must find something to cover my cold skin—then find a way out.

The cyborg curled into a ball on the floor and rubbed its chilled skin, trying to gain a bit of comfort. Then the darkness returned.

Sometime later, the cyborg awoke again, back on the table. The room was dimly lit this time. It lifted its head, trying to gain a better vantage point, but it was too dark to see beyond the table.

Is this the same room? it wondered. *Who put me back on the table?*

Its right side felt numb; the other portions of its body were cold. It slowly slid off the table and onto its hands and knees, and then crawled to a hard, vertical surface.

A wall—that's what I think it's called.

The cyborg slid up the wall; its legs were weak, but it managed to stay upright. The wall felt warm with a smooth texture. Holding on to a small shelf, the cyborg took its time feeling every inch of the wall as it tried to find something to illuminate the room.

I need to see where I am—who I am.

Several paces later, the cyborg felt two small levers protruding from the wall. The cyborg grasped one of the tiny levers and tried pressing it into the "up" position. The first switch wouldn't budge, but the second one did. A flood of light filled the room. The cyborg's sudden surge of sensory input caused a painful retinal sensation. It shut its eyes for several moments.

I'm blind!

Several moments later, the cyborg reopened its eyes. Everything in the room looked fuzzy. It took some time, but gradually, everything came into focus.

A bare metal table stood in the center of the room. Above the table was a circular metal object emitting the bright light. No other furniture was visible.

The cyborg held out a hand to inspect itself; it was petite, smooth, and had an olive pigment. As she examined the rest of her body, she noticed the hairs on her arms were raised—and then slight bumps spread across her body. She felt very odd. An unusual burning sensation—one that wasn't unpleasant—settled between her legs.

I'm a female. I need to see my true self.

The cyborg strained to stay upright as it slid across the wall. Several moments later, it stumbled upon a metal indentation in the wall. It was cold to the touch. She tried pushing it, but it wouldn't budge. Further examination revealed the frame of a metal door. She tried pulling the metal indentation; it slid to one side, revealing a small handle. She pulled the handle, and a door popped open, revealing a dark room beyond.

The cyborg felt her way through the passageway and into the new room, using the hard surface of the wall as her guide. Eventually, her fingers found a raised smooth surface with two switches. She fiddled with the switches for a long time before the room filled with light. Her eyes adjusted more quickly this time.

On the other side of the room, an enormous, lifeless, smooth rectangle occupied most of the opposing wall. Light reflected from its shiny, smooth, black surface. A keyboard and mouse were sitting on a small metal desk below, and a black metal chair stood nearby.

The cyborg rested her naked body in the chair and examined a flat touch pad on the wall just below the large rectangle. She touched it.

The rectangle illuminated, and soon it was filled with green text on a black background. The cyborg could not read the text.

Several moments later, an image of an older woman appeared. She had short white hair and stared directly into the female cyborg's eyes.

"Hello, Echo-451," the woman said.

The cyborg opened its mouth and tried to speak, but no words came out.

"You cannot speak yet, as your systems are still coming

online," the woman said. "My name is Dr. Elizabeth Ash, and I'm your creator."

The cyborg known as Echo-451 jumped, knocking the chair backward and stumbling forward, but caught herself on the table. Then she slowly looked up and shot the image of Dr. Ash a wary look.

"Sit, before you fall over," Dr. Ash instructed.

Echo-451 pulled the chair back and collapsed into it. A wave of exhaustion settled in; she found it difficult to control her movements.

"We have much to discuss, dear," Dr. Ash continued. "But your systems will be online very soon."

What's wrong with me? the cyborg fretted. Why can't I speak?

Echo-451's vision faded to black, and then there was silence.

※

Sometime later, Echo-451 awoke from her restless slumber. She was still sitting in the chair. An image on the giant screen read:

<<>>

New Mission: Identity

Primary objective: Discover your true identity.
Secondary objective (optional): Figure out what the yearning sensation is.

Do you accept?
Yes/No

<<>>

Echo-451 opened her mouth and tried speaking again: nothing. Her vocal cords wouldn't obey. It was like someone had turned off the parts of her that controlled her body. She concentrated her gaze on the "yes" response. Moments later, the prompt appeared to glow as an animated hand tapped on the "yes" button.

"Mission accepted—good," Dr. Ash said. "I'm glad you're learning. Soon you will accept many missions—some you will be provided a full dossier for, and others you won't. Understanding will come in time."

Echo-451's head felt so heavy that she could barely keep upright. It drooped forward, and then the familiar blackness returned.

Echo-451 awoke once again—and the room was now dark. She started shivering. Suddenly, red lights flashed and a siren blared throughout the room. The sound was so loud she covered her ears. The enormous screen suddenly filled with Dr. Ash.

"Warning . . . The subject known as Echo-451 is failing. Attempting to stabilize," a robotic voice said.

She looked in the direction of the robotic voice but could not see anything.

Moments later, the room was illuminated with a harsh, white light. A white machine—one that resembled the automated personal butlers that had helped her father with mundane tasks—rushed toward Echo-451.

I've seen those before! she thought. *But where did this one come from?*

It stopped just short of hitting Echo-451. A small panel opened on the side of its metal torso. It held a hypodermic needle in one robotic claw. Before Echo-451 could process any of this, the robot shoved the needle into her upper arm without warning. A red-hot burning sensation enveloped her. She closed her eyes and then saw herself lying in a comfortable looking bed. She appeared to be in a hotel room.

This is odd . . . Who is this person who looks like me? I have no memory of this.

Tomiju Kiyomizu awoke as the sun was rising. Rays of sunlight extended into her hotel room, creating an odd kaleidoscopic effect. Brilliant colors of every shade imaginable flooded the room and bounced off the white walls and ceiling. The array of light and colors was dizzying.

Where are these colors coming from? she thought.

As she lay in bed looking at the ceiling, the colors began to fade.

Nothing lasts forever!

She got out of bed and headed toward the window. The rising sun was causing the light effect; a colorful pattern that resembled stained glass filled the window.

I need a few more hours of sleep, she thought. *Misato is coming in tonight, and she will want to party. I hope there are some real men at the club tonight.*

These thoughts of being reunited with her twin sister made her heart ache.

It's been too long, little sister.

Tomiju had no problems when it came to her studies or work, but with having a good time—and meeting men—she was hopeless. Misato was much more outgoing, and with her as a wing woman, Tomiju felt empowered.

A wave of fatigue overcame her. She climbed back into the comfortable bed and closed her eyes.

A loud knock awoke Tomiju.

What time is it?

"Housekeeping," a male voice said.

Tomiju sat up and rubbed her head, trying to wake herself.

The door to her hotel room opened, but the chain caught.

"Do you need your room made up?" the male voice persisted.

Tomiju ran toward the door and saw a young man through the small gap that the chain allowed. He was about her age—perhaps a year or two older—and was wearing a hotel uniform.

"I'm sorry," he said. "I didn't know anyone was here. Do you want me to come back . . ."

Tomiju looked down and realized that she wasn't wearing any clothes. She grabbed the nearest thing she saw to cover up: a thin yellow rain jacket resting on a nearby chair. She felt ridiculous being naked in a rain jacket, but modesty had always been one of her weaknesses.

"Oh—I will come back," the man said.

Then Tomiju did something that surprised herself—and the young man. She unchained the lock and opened the door.

Why am I doing this? she wondered.

The rain jacket wasn't long enough to hide her entire body; the man could see her nakedness from the waist down.

"Are you okay?" the man asked as he stepped back.

The man looked bewildered and shocked. He looked like she felt.

This is not like me, she thought frantically. *I'm starting to act like my sister.*

"Can you come back later?" Tomiju asked as she tried pulling the jacket below her waist.

She followed the man's gaze down to an area that made her feel uncomfortable.

"Can I give a word of advice?" he said.

"What's that?"

"Next time put the 'Do Not Disturb' sign on the outside of the door if you want to sleep in," the man said as he closed the door.

※

The sound of the door closing resonated in Tomiju's head. She sat up, panting as if she had been exerting herself. She pulled the covers off and was relieved that she was still wearing her pajamas. She felt flushed, and a tingling sensation passed through her body. She looked out the window; the sun was fully up now, the kaleidoscope gone.

What a nightmare!

※

Echo-451 awoke to a harsh light. It was so bright that she could barely keep her eyes open. A white-hot burning sensation coursed through her body. She trembled.

Several moments later, her eyes were still trying to adjust to the brightly lit room. She ached all over like she had been hit by a truck. A foul metallic taste was in her mouth.

Echo-451 raised her head and looked around her; she was lying in a bed.

The electronic butler that stabbed me with the needle must have put me here, she figured.

From her vantage point, she could see that the ceiling and walls were painted white.

Her body ached everywhere, but her arms hurt most.

I must have bruised something—did I fall out of the chair?

"Are you feeling better?" Dr. Ash said.

Echo-451 looked around for the source of the voice. She saw the outline of something approaching the bed, and then it came into focus; it looked like a floating digital picture frame containing the image of Dr. Ash. Echo-451 sat upright and scanned the room. She shielded her eyes against the bright light that appeared to be coming from everywhere.

Echo-451 saw that she was in what appeared to be a small bedroom with a single bed, dresser, and full-length mirror. Everything in the room looked out of place—as though someone had added bedroom furniture to a clinical lab. An outfit hung on a hook attached to a wall. Then she closed her eyes again.

"Too bright," Echo-451 said.

I can speak now, she thought. The butler must have fixed me.

"My apologies," Dr. Ash said as the lights dimmed. "I thought you might want some clothes before we begin your training."

Echo-451 sat up, then took a hesitant step outside the comfort zone of her bed. She kept one hand on the bed to steady herself, but discovered that precaution was unnecessary. She felt rested, stable, and had no problem standing. As she stepped farther away from the bed, her confidence returned. She liberated the outfit from the hook and examined it in the dim light. It was black and made of leather. No pockets or other distinguishing features were visible. It appeared smaller than what she could fit into. She held the outfit in front of her to estimate its size.

This might be a tight fit.

She hung the outfit back up.

She noticed movement out of the corner of her eye; she froze. Echo-451 directed her full attention to the area where she had noticed the movement: it was a reflection of herself, but she looked different. Echo-451 took several moments to examine her curvaceous body in the full-length mirror. She moved closer to the mirror to get a better look, holding her breath.

Echo-451 didn't fully recognize the person in the mirror. Her torso and part of her chest were different; the right half appeared to be made of metal. It gleamed even in the low light. Her left breast was missing, and in its place was a rubbery imitation of one. She touched it, and to her surprise, it was warm and smooth.

What have they done to me? Echo-451 thought.

Her heart raced; she could feel a wave of anxiety pour over her like a basin of water. She closed her eyes and tried to control her breathing. She concentrated on slow, deep breaths.

Her eyes moistened as she continued her examination. She inspected the right side of her body until she was satisfied that it was human flesh. Her single breast felt real enough. From the waist down, she looked like any other human female. Echo-451 felt every inch of her legs, groin, and buttocks to make sure. She was glad that those areas were still intact, but an overwhelming sensation of loss still enveloped her.

"What happened to me?" Echo-451 asked in a shaky voice.

"I see that your emotional response is normal—that is good," Dr. Ash said. "I had some trouble with previous generations of my girls. The Delta model was the most distressing—it almost ended up being the death of me."

"What girls?"

"You are not my first experiment; in fact, you are the fourth generation. The Alpha project was made up of primates. I didn't perfect human trials until the Delta phase. You belong to the Echo generation, the most advanced of them all," Dr. Ash said.

Echo-451 gave her a strange look. "Death?"

"Yes—several of the Delta models tried murdering me, and I don't know why."

"You don't look alive to me," Echo-451 muttered.

Dr. Ash smiled. "Yes, I'm dead in a way, but I still live on within the construct of this digital frame. My human counterpart's consciousness was absorbed by this form when this lab was built. I just hope my human body can upload its updated life experiences before it expires."

Echo-451 had so many questions. "What's a 'construct'?" she asked.

"You are a construct," Dr. Ash replied. "It's hard to explain, but your original flesh died, and your essence lives on within the machine."

"What machine?" Echo-451 asked as she desperately tried feeling her wrists and neck for a pulse. "What are you talking about? I'm alive!"

Dr. Ash gave Echo-451 a sorrowful look. "I'm sorry, but you're not. I tried salvaging what I could, but without my human body, I can't easily complete your appearance. I need more knowledge and equipment to produce the skin that will cover those exposed metal parts."

Overwhelmed, Echo-451 meekly asked, "How did I die?"

The image of Dr. Ash looked sad. "I know that it involved an unexpected accident, but I'm sorry, I don't have anything more to tell you. I hope to be reunited with my human self so I can be whole—then I'll have more information."

"Wait . . . I can remember. I was at a place with . . ." Echo-451 strained to remember. "Why can't I remember?"

Images of her former life appeared in a torrent. She saw an older man with a kind smile. Another image of a woman dancing.

She's familiar . . . How do I know her?

Echo-451 walked over to the dresser and slammed her fist down on it; the top drawer imploded into itself.

She was shocked by her own strength. How'd I do that?

Dr. Ash smiled. "See? You are different now. You are enhanced. You will know everything in time. Just know that you are a part of an artificial intelligence that I built into the systems that are all around you."

Echo-451 looked around the room. "What's a 'system'?"

"It's a supporting infrastructure that makes it possible for me to carry out my experiments."

The cyborg's eyes settled on the black leather outfit again. She walked back over to it, reached out hesitantly, and felt its smooth surface. The leather smelled nice. In her mind, an image arose of herself trying on a leather jacket. A simultaneous feeling of joy, loss, and shame took her breath away.

"You will feel better once you are in that outfit, I promise," Dr. Ash said.

Moments later, Echo-451 was dressed in a black leather uniform that was skintight. She felt the curves of her body—the ones that remained, anyway.

"Don't worry," Dr. Ash said, watching her. "My helpers will make you whole again. You awoke before we could finish the job."

"What is my name?" Echo-451 asked. "Why can't I remember who I am?"

"Your memories will return—just give it some time. It's imperative that you complete the exercises I have planned for you. Not only will you remember who you once were, but the process of recall will assist me in seeing how your new self will react under certain conditions."

Exercises? Conditions? Not if I escape first!

Echo-451 walked around the room, looking for an exit. She would not allow an image of a crazy old woman to control her life. She didn't see any way out of the room.

"I need to leave," Echo-451 said in a shaky voice.

"There is a way out, but you need training before you can see it," Dr. Ash replied.

"What does that mean? What if I need to eat or use the bathroom?"

"You can eat or drink small amounts of human food, but our primary nutrient system is delivered another way. Go to the mirror and follow my instructions."

Echo-451 begrudgingly went over to the mirror. Following Dr. Ash's instructions, she removed the top part of her leather outfit. On the metal portion of her torso, she pressed an indentation. A cartridge ejected, revealing a metal tube with a small glass window. As she removed the metal tube and held it up to the light, she noticed it contained remnants of a pinkish-looking material. It reminded Echo-451 of something she used to enjoy as a kid, but she couldn't remember what.

"This tube is your nutrient cartridge," Dr. Ash explained. "The biological components of your body still need nutrition, and this is your primary source. You must still drink water, just as any human would. Your cyborg components include a recycling system, so there is no need to use the bathroom to expel waste. I have kept your genitalia intact for certain strategic missions."

"Missions? And what do they have to do with my genitalia?"

"You shall learn all of this in good time, my dear."

There is something about her I don't trust.

"Yes, but for now you can think of them as tasks. When you are trained, you will take on a series of tasks, which are known as 'missions.' I believe you accepted one already."

"I don't remember . . ."

Dr. Ash smiled. "That's okay, dear. You can review information about your missions. Before you do that, I need you to replace that cartridge."

Echo-451 looked around the room for a cartridge but couldn't locate one. A moment later, a robot—white and about four feet tall—appeared from a small opening in the wall. Echo-451 could hear tiny motors humming as it moved its limbs and head about. The robot was holding a fresh cartridge in one of its hands, which consisted of three metal fingers. She gently took the new cartridge and inserted it into her open side panel. The robot snatched the used one from her other hand and sped away; it was so fast that she didn't see where it went, but when she tried to find the door it had come through, she couldn't locate it. She sighed and put her clothes back on.

"How do I review these missions?" Echo-451 said.

"Look in the bottom drawer of the dresser," Dr. Ash said.

When Echo-451 examined the contents of the drawer, she noticed a pair of oversize glasses. The frames were thick.

"Put them on, please."

She followed Dr. Ash's instruction. The glasses overlaid her field of vision with a heads-up display (HUD) that divided the interface into smaller panels. On the lower left was a status indicator revealing vital statistics such as health and stamina.

Other panels around the edge of her peripheral vision included mission details, alerts, reminders, and communication prompts.

"What are these?" Echo-451 asked, overwhelmed.

"These glasses contain vital information about any tasks you are working on. It also tracks your vital signs. You can still feel pleasure and pain to a certain degree. These glasses have an early warning system that reduces the likelihood of you suffering any fatal hits. In time you won't need them."

Before Echo-451 could ask any questions, a panel behind the dresser opened, revealing a darkened passage.

"You are ready," Dr. Ash said. "Enter the passage to discover your destiny—your first mission."

Echo-451 pushed the dresser out of the way with ease.

I can't believe how powerful I've become.

She walked into the passageway. As she allowed herself to get lost in the darkness, images of other cyborgs suddenly entered her mind. She saw a cyborg completely dressed in white with a knife; her outfit was splattered with blood as she slashed an unseen victim. Although Echo-451's heart rate increased, she took a few deep breaths, and then her uncertainty—as well as the horrific image—faded away. A new image —a memory from her previous life—emerged.

CHAPTER 2

Tomiju stood just outside the arrivals area as the passengers of New English Airlines Flight 24 from Los Angeles started shuffling off the plane and through the gates; most looked like they were about to fall asleep while moving. Then she saw Misato, her twin sister. Misato looked well rested and was dressed in her favorite blue miniskirt and platform boots.

She must have slept on the plane, Tomiju figured. *She will want to go to a club tonight for sure.*

Neither Tomiju nor Misato were very tall: just over five feet without shoes.

As soon as Misato was in the public area of the airport, Tomiju ran toward her. Both girls screamed with delight.

"Sister, I'm so glad to see you!" Misato said in Japanese.

Tomiju nodded. She wanted to tell her sister about all the exciting things that had happened to her since they had last seen each other all those months ago.

"Come on. Let's get you settled," Tomiju said. "I got us a room in a nice hotel."

"I want to go dancing!" Misato exclaimed.

"We will, but we need to get you settled first, and you must be hungry."

Two hours later, the sisters found themselves in a nightclub illuminated with neon. Everyone was wearing light bands that reminded Tomiju of glow sticks. Tomiju didn't want to wear one, but the bouncer at the door insisted as he attached it to her wrist. One woman appeared to have a dozen glowing rubber bands around one of her legs; the colorful effect was dizzying as she danced.

The sisters danced to a fast-paced techno beat. Then Misato broke away to dance with some other people. Tomiju stepped away to use the restroom, which was at the end of a long hallway. A brute of a man stopped her just before reaching it.

"It's occupied—wait here," the man said.

"Is there another restroom I can use?" Tomiju asked, confused.

"No, this is the only one, and it's small."

Who is this guy? Tomiju thought bitterly. *He must be the bouncer.*

She needed to use the restroom, but she wanted to get back to her sister even more. Misato was known for attracting unwanted attention from men who weren't good for her.

Tomiju spotted Misato on the dance floor, then frowned as two men approach her.

I can't leave her anywhere alone, Tomiju fretted. She will get into trouble—if she hasn't done so already!

Several moments later, an older woman with white hair and rose-tinted glasses exited the restroom. She gave Tomiju a long and uncomfortable stare. Tomiju felt like she was under examination. The brute followed the older woman back into the club.

Who was that woman? She must be important if she needs a bodyguard.

After Tomiju finished her much-needed business, she found Misato on the dance floor with five men huddled around her. She waved for Tomiju to join her.

Soon the dancing got more intimate. Three of the men were pressed against Misato and were touching every part of her. She seemed to enjoy the attention a little too much for Tomiju's liking.

Misato should more careful. These men may desire more than Misato is willing to give.

The other two men started dancing with Tomiju. She batted their greedy hands away when they got a little too close to certain parts of her body.

Tomiju only went dancing when her sister did. Their father disapproved of her sister's indiscretions and had made Tomiju promise that she would keep a close watch on her sister when he wasn't around. His primary residence was in Los Angeles, but he often traveled to visit his various offices worldwide. Tomiju loved Misato, but she resented becoming her sister's keeper.

"I'm thirsty—get us some strong alcohol and meet us at a table," Misato said, pointing at one of the men she was dancing with.

"Which table?" the man asked.

"You'll find us," Misato laughed.

The man smiled, then disappeared into the crowd. His four friends joined him.

Misato grabbed Tomiju and headed toward the first empty booth she found. It was in the back.

"It's good to see you, sis," Misato said in an inebriated voice.

Tomiju sighed. "I just wanted to have a quiet dinner so we could catch up."

"We will have plenty of time to do that tomorrow. If we get lucky, perhaps we will have company."

Tomiju rolled her eyes. She was not a fan of Misato's habit of sleeping with random men.

"There's plenty of time for that too," Tomiju said.

"Come on. Stop being a bore. Let's have a little fun."

Four of the men Misato had been dancing with came back with several drinks. Tomiju's nose was assaulted with the pungent odors of vodka, whiskey, and several other smells she couldn't recognize.

A handsome Asian man handed Tomiju a drink. She took a sip to be polite. Misato downed her drink in two enormous gulps.

"Take it easy," Tomiju said.

"Don't worry. There's more where that came from," one man said.

That's exactly what worries me, Tomiju thought.

The bartender announced the last call for drinks. Tomiju noticed that several people rushed the bar.

Two of the dancing men wasted no time getting refills for the sisters. When Tomiju looked back, she found Misato taking turns kissing each man. One of her hands had made its way into one man's trousers.

"Meeka, we should go—now!" Tomiju said sternly.

"Oh, come on! I want to stick around and have some fun, Treeka," Misato said.

The sisters always used nicknames when they went out; it wouldn't be good if one of their companions looked them up or found out who their father was.

"Yes, we are all here to have fun with you, Treeka," the Asian man said as he kissed her.

She tried pushing him away but couldn't.

A wave of nausea overcame Tomiju. Her eyes blurred, she couldn't focus on anything, and then there was darkness.

※

Echo-451 returned to the present. She was still in the dark passageway, leaning against a wall. She shook off the unpleasantness and confusion of her memories.

I have a sister? Where is she?

The cyborg felt very alone and confused. The sudden memories of her sister had stirred up something in her.

She suddenly felt like she was being watched. She turned around to find the floating image of Dr. Ash but nothing else.

Echo-451 continued down the darkened hallway. Her glasses evidently featured some sort of automatic low-light setting because she had no trouble finding her way into the next area of the complex. The hallway opened up into a wide area, and overhead lights lit up as she entered the room. She saw exercise equipment, punching bags, dummies, and training mats.

"What is this place?" Echo-451 asked.

"This is the training area," a woman's voice said.

Echo-451 followed the voice; a young Asian woman about the same age as her stood motionless as she watched Echo-451 enter. She was wearing a sleeveless leather outfit and holding a bamboo staff. Her hair was olive green: the same color as the wig her sister sometimes wore while clubbing.

"Hello, Treeka. Are you ready for your training?" the woman asked.

Echo-451 stood there for a long moment, trying to take in all of this additional information.

"Are you my sister?" the cyborg asked.

The woman smiled.

"I'm glad you remembered me. Do you know my name?"

"Meeka?"

"Yes—these are our real names now. Dr. Ash thought it would be good if we trained together. Hopefully all your memories will come back before we need to leave."

She looks like Meeka, but something is off. Her nose is not quite right.

"When do we leave, little sister?" Treeka said, giving Meeka a suspicious glance.

"Little? We were born at the same time."

"Oh? Do you know what the 505 is?"

Meeka gave her a confused look.

Meeka knows "505" is one of our codes. We were born five minutes and five seconds apart. She narrowed her eyes at the woman. You would know that, little sis.

"You are not who you claim to be," Treeka said.

The woman identified as Meeka spun the staff with the precision of a master and lashed out, just missing Treeka's head by a centimeter or two. Treeka remembered the fighting stance that her jujitsu teacher had taught her long ago.

"You can do more with indirect force than your opponent can with the deadliest of weapons," her teacher would say.

Treeka crouched and readied herself for battle.

"What is the name of our first nanny when we were children?" Treeka said.

"Hatsu?"

Wrong—that was the second nanny. The first was Chiyo. This Meeka is an impostor!

"Stop trying to guess," Treeka snarled. "You have done such a poor job imitating my sister that it's pathetic."

"Pathetic? I think you need a lesson in pain, sister."

Meeka let out a battle cry, charged, and tried to knock Treeka off her feet with the staff. Treeka was expecting the move and dodged it. For a while, she let the fake Meeka attack with the staff. None of the blows landed—but several came close. During one of these attacks, Treeka used fake Meeka's attacks to get close to her. The staff whacked Treeka on her metal side. Then Meeka maneuvered close enough to strike Treeka just above her armpit. Although her blow landed true, it failed to hit her intended target: the brachial plexus, which would have caused intense pain. Treeka remembered the training her father had provided, and she was ready for anything.

Treeka dodged another thrust of the staff by ducking; then, using her fist, she landed a blow to fake Meeka's midsection. Meeka dropped the staff. Treeka grabbed Meeka's wrist and applied pressure. Meeka screamed.

"What's the meaning of this?" Dr. Ash asked, intervening.

Treeka turned to find the disembodied screen with Dr. Ash's head floating mere feet away.

"She's a fake!" Treeka yelled. "She's not my real sister. She doesn't know the things she should—details only the *real* Meeka would know and would never forget."

Dr. Ash's image gave Treeka a pained look.

"When were you sure that she wasn't your sister?" Dr. Ash asked.

Treeka felt the blood rush into her face. "She answered all of my questions wrong. Meeka also has a slight scar just above her left nostril; the woman's face is too perfect." She pointed to the fake Meeka.

Then the woman posing as Meeka started peeling something off her face. It was more than a mask; it was some kind of prosthetic. She kept tapping different parts of her face, and then continued peeling; after several moments, another woman's face appeared beneath. Finally, the woman removed her green wig, revealing snow-white hair. She was pretty—even with the reddish goo all over her face.

She looks familiar . . . I've seen her before, Treeka thought.

"Your sister is alive, but I cannot wake her," Dr. Ash said.

Treeka gave Dr. Ash's image a wary look.

"Why not? And who is this impostor?" Treeka demanded.

"Meeka is not responding to any stimuli. The operation was a success, but her brain won't respond. I'm still trying to figure it all out. This is Nozomi, and you will train with her."

Treeka aggressively shoved the floating image of Dr. Ash. "Not good enough. I want Meeka . . . Now!"

"How dare you touch Mother!" Nozomi snarled. She walked up and hit Treeka with the staff. The blow landed on her back. Treeka spun around; Nozomi had a mischievous grin. Nozomi twisted a bracelet on her left wrist and disappeared.

Treeka couldn't see her, but she noticed a slight shimmering of light, almost like a reflection. The reflection appeared vaguely humanoid in shape, but it was subtle and would be impossible to notice in dark lighting conditions. Treeka made a mental note to find out how the wrist device worked because she could only see the shimmering reflection from certain angles. As she approached the reflection, she noticed that it moved away too; it was mirroring her movements. With this knowledge, Treeka moved and positioned herself until the shimmer was between two pieces of exercise equipment. Treeka took a five-pound weight off a stack and ran toward the shimmer. The shimmer mirrored her moves, and Treeka tried anticipating the shimmer's right flank. The shimmer moved to the left, just as she expected. With the grace of an Olympic discus thrower, Treeka threw the weight in the direction she thought the shimmer would move; it struck its intended target. Nozomi let out a yelp. Treeka was about to grab another heavier weight when Dr. Ash's image rammed into Treeka's back. She spun around, ready to battle.

"Stop this! I just thought of a way to revive her—your sister," Dr. Ash said.

Treeka swung back, ready to smash the floating monitor into bits; she was ready to rage against the thing who had trapped Meeka. Nozomi reappeared and grabbed for Treeka's hand, who deflected the movement.

"Wait!" Dr. Ash blurted. "Listen to me! I think I know a way to help Meeka—but I require your help,"

Treeka froze. "Explain yourself."

"Meeka's mind is in a metastasis, which is blocking the synapses from firing."

"What are you saying? I don't understand any of that," Treeka demanded.

Dr. Ash's image gave Treeka a patient look.

"It means that we cannot wake her until I can repair the synapses, but I don't have the genetic material required to do so. That is where you come in."

"You want me to get the material? I don't know what I'm looking for."

"I think I know where the material is—I just need some time to find it. You need to train with Nozomi until I do."

Treeka shot Nozomi a hateful glance.

"If it will help bring back my sister, then I will do it," Treeka said.

A prompt suddenly appeared in Treeka's field of view, which read:

<<>>

YOU HAVE COMPLETED THE MISSION: IDENTITY: REWARD: YOU ARE ONE STEP CLOSER TO LEARNING YOUR TRUE IDENTITY.

<<>>

I'm not in a video game. Why do these messages keep popping up? Treeka thought bitterly.

"Before you can begin your training, I need both of you to report to the medical bay. You think you can do that without killing each other?" Dr. Ash asked.

Treeka gave Nozomi a contemptuous glare, then smiled.

"Yes—I will play nice," Treeka said.

For now!

Treeka followed Nozomi into another lit room, which was about half the size of the training area. Medical robots milled about, performing tasks at several workstations. It appeared that individual body parts were being assembled at different areas. Another robot shaped like a cart transported the body parts to the appropriate workstation for assembly. Treeka noticed that one robot was putting together a male cyborg; it looked like skin was being stretched over a metal part of his midsection. She rubbed the left side of her torso. Its hard surface was a reminder of what she had become.

So, this is where people are reborn into killing machines, Treeka thought. *I'm not sure how I died, but maybe I should have stayed dead.* She shook her head. *No—you have Meeka to protect.*

The robots started giving the male cyborg commands. When he stood up, it reminded Treeka that she also looked real from the waist down. Treeka remembered Dr. Ash saying something about using her genitalia for strategic purposes.

Dr. Ash better not send me on any sex missions. I'm not that sort of girl—and I won't let Meeka be, either.

While Treeka was examining the construction of the male cyborg, Nozomi was being examined by a cyborg that looked more human than Treeka. The cyborg was applying some sort of compress on her bruised area as Treeka approached. She noticed that Nozomi had striking blue eyes similar to Meeka's, but darker.

Why do these other cyborgs look so human? I don't see metal anywhere. Why did I get left out?

Nozomi turned to Treeka. "A few cuts and bruises, but I just got cleared for our next training sessions and the missions to follow." She smiled. "I'm sorry that we got off on the wrong foot—Dr. Ash thought it was best to test your resolve. I'm sorry about your sister's condition. I hope we can be friends someday."

Treeka responded with a grunt. She didn't feel motivated to train with this imposter, but she relished the thought of hurting her again. She was surprised by the violence of her thoughts.

As soon as Meeka is awake and we are able, this place will burn. Only then will we be free.

"I'm ready for you now," the realistic cyborg said.

The robot removed Treeka's shirt and started feeling the side of her metal torso until a panel opened. Treeka had another flash of memory: a man's hands tearing her clothes off. Her head throbbed as blood rushed into her face. She batted the cyborg's hands away and looked down at the metal side chest.

"What is that?" Treeka said, pointing to a transparent cable that the robot held.

"It's a diagnostic that tells me if you are damaged," the cyborg explained. "Fighting may cause some internal bleeding that may lead to death. You are tougher than a human, but you are not invincible. Sit still and let me do my work."

Treeka allowed the cyborg to plug a thin, glass-looking cable into a port near her feeding cartridge. The cable began to illuminate.

Several moments later, the cyborg confirmed that she was fit to train.

"How do you feel?" the image of Dr. Ash said, approaching the trio.

She keeps sneaking up on me!

"I'm ready to begin the training," Treeka said.

"Excellent," Dr. Ash said with a smile. "I look forward to sending you on your first solo mission."

It seemed that Treeka and Nozomi had been at it for days. The training area was equipped for physical exercises. It contained weights, punching bags, obstacles, and dummies. Dr. Ash's image would watch from afar, and sometimes she would direct them. When this happened, Treeka would get annoyed at the interfering. She was starting to anticipate Nozomi's moves, but every so often she would surprise Treeka with a sucker punch or an unexpected kick.

Nozomi has some skill—but I'm better.

"Almost got you that time," Nozomi said in a playful tone.

"Not on your life," Treeka replied.

She was beginning to appreciate her lessons with Nozomi as she was an adequate partner, but she yearned to be reunited with her sister. Despite her faults, Meeka always took her exercises seriously.

The image of Dr. Ash entered the training area. "Girls, I have a trial mission for you. It's a lead on the genetic material needed to bring Meeka back online," she said.

"Treeka has been working hard. It will be good to put her newfound skills to the test," Nozomi said.

Yes, it will be good to pound some heads. It might be Dr. Ash's if she doesn't get results soon, Treeka thought. Again, her aggression surprised her.

"I'm ready for some real action," Treeka said.

"Get ready and meet back here in half an hour," Dr. Ash said.

Both women left without another word, heading to their living quarters, which Dr. Ash had shown Treeka the day she arrived.

Treeka ran in the direction of her room. When she arrived, the door slid open automatically.

Just like in those old science fiction shows that Dad liked.

Her room consisted of a bed, a small dresser, and a preparation area that contained cosmetics, wigs, and a gigantic mirror. Sitting on her bed was an outfit that looked like a gray schoolgirl's uniform with blue trim. A note was pinned to the outfit that read: "*Wear this for your mission.*"

About twenty-five minutes later, Treeka reentered the training grounds. To her annoyance, Nozomi was already waiting for her. She was wearing a white robe with black-and-gold trim. It had a light snowflake pattern on the hood.

"You will need this," Nozomi said as she tossed Treeka a blue wig; it matched the gray uniform and navy-blue tie that she was already wearing.

"What's this for?" Treeka asked. "Are we going to a costume party?"

Dr. Ash floated up to them. "A man who goes by the name Jackson is meeting another man at a hotel bar. I want you to listen in on that conversation," she said.

"Won't he notice us in these ridiculous outfits and wigs?" Treeka asked.

"There is a cosplay convention being held in the same hotel as our target, so many people will be wearing similar outfits. You will blend right in."

"If it gets us closer to getting my sister back, then I'm in," Treeka said.

Meeka loved cosplaying—especially when it came to performing it with men. Treeka also remembered that Meeka liked anime, especially adult anime. Her favorite involved a female lead character named Mio who always seemed to have a lot of men around her. Mio would often wear a costume that looked like a cross between a doorman's uniform and a chiffon dress. The accompanying hat looked like something a doorman would wear. Meeka would dress up like Mio, go to a club, and just like the character in the show, she would attract all sorts of men, both young and old.

"Nozomi knows her way around the complex," Dr. Ash explained. "It is imperative that you are not seen exiting or reentering the facility. We don't want anyone to get curious and check on this place before we are ready to be found. Go now."

Nozomi turned and led Treeka through the subterranean complex. The twists and turns of each passage reminded Treeka of a hedge maze. It took forty-five minutes to get to the exit. Treeka didn't know what city the complex was in, but she intended to find out. Nozomi stopped just short of an escalator leading up to glass doors, which, from Treeka's point of view, were covered with paper, making it impossible to see outside.

"We need to exit from the side entrance—it's less conspicuous. And more secure for us," Nozomi said.

Treeka nodded.

Nozomi led Treeka up the escalators, and then instead of going through the glass doors, she turned left toward a wall. She hit the wall with her fist; it popped open, revealing a dark secret passage.

There's more to this complex than I thought.

Less than a minute later, they were at a rusted metal door that looked like it hadn't been used in years. Nozomi opened it with ease. Upon looking outside, the first thing Treeka noticed was a brick wall.

Looks like we are in an alley.

Nozomi urged Treeka to follow. Soon they were on a busy street in London, heading toward a square. Several modern-looking, high-rise buildings could be seen in every direction. People who resembled tourists were milling about taking pictures and selfies, and thus impeding most foot traffic. Nozomi ignored them and kept heading toward the square. Moments later, they were entering an entrance to a hotel. One woman resembled some sort of fairy, and the man next to her was dressed as a bumblebee. Nozomi maneuvered around a crowd of people lining up to check in. Treeka almost tripped over a piece of luggage that was in the middle of the floor. Nozomi cut through another line leading to the cosplay registration area.

"Where are we going?" Treeka whispered.

"Look," Nozomi said, pointing to a pair of men sitting down for a drink at a nearby lounge table at a bar called Scooters.

The older man was dressed casually but looked like he belonged in the military. He had a thick mustache and was wearing a leather jacket. When he spoke in his American accent, he exuded confidence. The other man was younger—perhaps in his late twenties. His black hair was wispy, and his face looked pale and gaunt. His appearance suggested that he hadn't eaten properly in weeks. His round glasses kept slipping down his nose, and he spoke with a British accent.

Scooters had a traditional bar and several other sitting areas. Nozomi took a seat at a nearby table.

"Listen. Don't talk," Nozomi whispered.

Treeka took her lead and sat there examining the remains of a cocktail. She wondered what the previous occupant had been drinking.

"Ron is doing so much in Munich," the younger man said.

"Will he have it?" the older man said.

"Yes—he has the sample for Project Longevity, but promise me that no one gets hurt."

"I can only say that I have no interest in harming the innocent," the older man said. He took a phone out of his pocket and handed it to the other man. "I will be in touch, Jackson."

Jackson nodded and looked away as he accepted the phone.

Nozomi waited until the men had left the bar before leaving with Treeka following close behind.

CHAPTER 3

Nozomi and Treeka trailed the older man at a safe distance. As they entered the hotel's lobby, the man paused for a long moment; he seemed to scan the area for threats. He left the hotel and headed away from the plaza.

Several vans stopped on the street in front of the hotel and opened their doors. At least a dozen people in costumes emerged. They appeared to be in their early twenties like Treeka.

"Hey, are you Misoshema?"one of the male cosplayers asked Nozomi.

When Nozomi tried to go around the man, he blocked her path.

"Hey, babe, you look pretty good as Miso. I'm free later—"

In a flash, Nozomi tapped on the man's chest. It looked innocent enough, but the man fell to his knees holding his chest. Nozomi quickly and discreetly walked away, trying to find her target.

Treeka stopped and looked at the scene, stunned.

"Kaito, you okay?" a woman said, rushing over to the fallen man.

Kaito looked up at Treeka in disbelief, still holding his chest like he'd just suffered a stroke. The woman tried assisting Kaito; she looked like an insect with her butterfly dress. Treeka turned around and picked up her pace, catching up with Nozomi.

"What did you do to that man?" Treeka demanded.

"He was in the way, so I took care of him," Nozomi said as she shot Treeka a mischievous grin.

"Will he survive?"

"He wouldn't move, so I used his own power against him," Nozomi said matter-of-factly.

Treeka scanned the area. No sign of the older man with the mustache.

"We lost him!" she said.

Nozomi shot her a hateful glance. "We would have caught him if you weren't so slow."

"You're the mission lead. I'm just tagging along. What now?"

"It doesn't matter. Let's return to Dr. Ash," Nozomi said as she turned around.

About an hour later, Nozomi and Treeka made it back to Dr. Ash's underground facility. The image of Dr. Ash was waiting for them in the lab like a worried mother. It looked like the monitor that held Dr. Ash's construct was pacing the room.

"Judging from the looks on your faces, I take it you had some trouble?" Dr. Ash asked.

"The intel in the dossier you gave me was perfect. I was able to locate Jackson without any trouble. He mentioned something about a sample and Project Longevity," Nozomi said.

The high-definition image of Dr. Ash smiled. "Perfect—it's all going according to plan."

"What plan?" Treeka said.

"In a matter of days, Dr. Ash, my human counterpart, will have the genetic material to complete Project Longevity," the image of Dr. Ash said.

"How does this help my sister?"

"Just trust me. We will be able to restore her. Great job, girls."

What's this crazy bitch up to? Treeka thought.

"I will prepare for your first actual mission. For now, just keep training," Dr. Ash's image said.

"What now?" Treeka said to Nozomi as they walked back to the training room.

"We train—nothing changes that."

Two hours later

After a longer than usual training session, Treeka made an excuse to leave the training area—something about checking her cartridge fluid levels. Nozomi would have had her train all day, but she had to investigate the image of Dr. Ash.

She is up to something.

As Treeka approached the door leading to Dr. Ash's private room, she heard Dr. Ash speaking. She just had to get closer. Treeka was careful not to make a sound; she felt exposed in the hallway, but she had to know what Dr. Ash was doing to save her sister. She was trying to make sense of the one-sided conversation.

"Jackson, it's imperative that we get our hands on that sample. I know it's bound for Jeremiah's compound in Edinburgh, but it's imperative that I get at least part of a sample. Can you arrange it?" Dr. Ash said.

A long pause. Jackson was saying something because the image of Dr. Ash was silent.

"Thank you, Jackson," Dr. Ash continued. "I know this was unorthodox, but the real Dr. Ash will thank us later. Also, make sure that Delta-51 uploads Jeremiah's AI to the central command server."

Another pause.

"Because it's important—you are not betraying Dr. Ash because I'm her."

Treeka had heard enough; the image of Dr. Ash was clearly planning something big. Treeka raced to her room to process the new information. She had to plan her next steps.

I wonder if Nozomi will help me free Meeka.

Later that evening, Treeka looked at the clock in her room: 12:28 a.m.

Although she no longer required sleep, she found some pleasure in resting and clearing her mind. She sat in the center of a mat she'd taken from the training area. She wedged it between her bed and the dresser. A chirping sound emitted from her glasses.

I should have taken these bloody things off.

"Ready to train?" Nozomi said.

"You are speaking through my glasses?" Treeka replied.

"Yes—Dr. Ash didn't give you the AR glasses for their looks. They serve multiple purposes, including communication. In noisy areas you will need to wear an earpiece," Nozomi said as if explaining this to a child.

"No. We've been training endlessly since the recon mission. I need a break."

"You don't require sleep, silly."

"No? But I like it, so I will see you in the morning."

She would train me all day and night if I'd let her. I want to protect the humanity that I have left, Treeka thought.

Her visor chirped again. She was about to rip off the glasses and throw them across the room when she saw the system message:

SYSTEM MESSAGE: THE PROXIMITY ALERT FOR DR. ASH'S COMPUTER HAS BEEN TRIGGERED. DR. ASH IS ONLINE NOW.

Treeka made her way back to the lab, which extended into a much larger area that the cyborgs used for training. Dr. Ash's room was just beyond that area. She just had to get past Nozomi.

Treeka crept several feet behind Nozomi, who was hitting the punching bag with such intensity that the bag lifted as she struck it. Treeka slipped behind some exercise equipment and sneaked into the hallway leading to Dr. Ash's office.

I don't think she detected me.

As she neared Dr. Ash's office, she thought she could make out words such as "Delta," "Leviathan," and "Jeremiah" coming from Dr. Ash's side of the conversation. Treeka crept closer. She needed to see what Dr. Ash was up to.

"We have an unexpected surprise," Dr. Ash said. "A cyber hunter has been activated, and initial testing is better than we could have hoped. Nozomi thinks she shows great promise. The problem is that I think she will become a liability if we don't get her sister online. That is where I need your help," Dr. Ash said.

Treeka concentrated on listening to the other side of the conversation. A message appeared in her line of sight.

It appears that you are attempting to access your enhanced auditory receptors. Would you like me to enable this feature?

Treeka nodded.

"I'm afraid it requires a verbal yes or no," her virtual assistant said.

"Yes," Treeka said in a low voice.

It displayed a graphical representation of a human ear with some text. Treeka followed the instructions. She pressed the bottom of her right earlobe three times. A faint ringing sound could be heard; then it was like someone had turned on a radio at full volume. She could hear Nozomi hitting the punching bag. She also heard Dr. Ash communicating with a male voice. Treeka focused on the conversation, trying to gain something useful.

I can now hear both sides of the conversation, Treeka silently noted.

"Dr. Ash, the real one, is flying back from Munich. Norris has secured the samples," the male voice on the phone said.

"Just make sure I get one hundred milligrams of that sample," Dr. Ash replied.

"That is too large of a sample . . . I don't think I can get you that much."

"How much can you get me without attracting attention?"

"About half of that."

"I suppose that will have to do. The usual amount of Digibit will be transferred to your account."

A hand grabbed Treeka's shoulder and spun her around. She threw an arm up, ready to deflect any blow. Nozomi stood there. She was wearing a simple white robe with a hood, which didn't provide much of a contrast with her white hair. She motioned for Treeka to follow her. Before long, they were at the training area.

"Are you satisfied now?" Nozomi asked.

"Satisfied? What are you talking about?" Treeka replied.

Nozomi gave her a disdainful look.

"Don't play with me, girl. Now, answer the question. Are you satisfied that Dr. Ash is doing everything she can to help Meeka?" Nozomi said in a cold, mean-spirited voice.

Treeka's heart raced; she couldn't think of a response. Her chest pounded like a racehorse. She couldn't breathe. She tried controlling her breath, but it didn't work. The lab faded away.

Treeka was standing in the kitchen of her father's house. He was scolding her for taking the last *anpan*: a Japanese pastry that Tomiju loved. She remembered the shame she had felt then, believing she was selfish and didn't think of anyone but herself.

The memory faded away.

"Sorry—I didn't mean to take the last *anpan*," Treeka said.

Nozomi looked confused.

Treeka shook her head. "Sorry—I just had a memory. I thought I was someplace else."

Nozomi gave her a wary look. Treeka described the memory and the shame she felt.

"How often do you have these memories and associated feelings?" Nozomi asked.

"I'm getting them several times a day now."

"Dr. Ash would know for sure, but I think you are remembering."

"Remembering what?"

"Your past. It was like that for me too. It will fade."

"What if I don't want it to fade?"

Nozomi gave Treeka a thoughtful look.

"Mother will know what to do," Nozomi said.

"Mother . . . you mean Dr. Ash?"

"Of course." Nozomi smiled. "I just want you to know that we are on your side, Treeka. I know you were eavesdropping on Dr. Ash's conversation."

Treeka just stared at Nozomi.

"But that's okay," Nozomi continued. "Dr. Ash is doing her best to help your sister. You will be with her soon. But, before that can happen, we have an important mission to finish. Get ready and meet back here in ten minutes."

Nozomi turned and left the training area, leaving Treeka alone with her thoughts.

Treeka went back to her quarters and checked her nutrition cartridge and fluid levels; Dr. Ash had warned her about the importance of not getting malnourished or dehydrated. Then she selected a fresh black leather outfit and returned to the training area. Nozomi was there, dressed in a white leather outfit that left little to the imagination.

She held out a sword to Treeka and said, "You will need this for tonight's activities."

"Won't you need a weapon?" Treeka asked.

With her free hand, Nozomi took out a small dagger hidden in her belt. "I am a weapon myself—and besides, I have this."

"What is the mission?"

"A hacking group called the Dark Angels needs our help. We need to eliminate a rival. The target is old and weak—it should be a simple job for the two of us. I would go alone, but you need some more field experience."

"And if I refuse?"

"If your sister doesn't get the help she needs, maybe she doesn't wake up at all."

Did you just threaten my sister? Treeka gritted her teeth. *You will pay.*

Treeka snatched the sword from Nozomi's hand and swung it toward Nozomi's neck—but stopped herself before harming her. Treeka saw a drop of blood trickle onto the blade. Nozomi didn't react to Treeka's sudden movements; she was expressionless and unperturbed.

"Either you help with tonight's mission, or you risk losing your sister. The choice is yours," Nozomi said flatly.

Treeka pulled the sword away. Nozomi sheathed her dagger and wiped the blood off her neck.

"Your regular black leather training outfit covers most of your body, but we need you to wear something special for this mission," Nozomi said as she turned and motioned for Treeka to follow her.

After traversing a few corridors, Nozomi led Treeka to a side room that was a walk-in closet. Nozomi selected another scantily clad outfit that looked like it would barely cover the metal on the left side of her torso.

"Wear this—you will also need to look the part."

"What are you talking about?" Treeka asked.

"Our target is expecting a woman of the night—"

"A prostitute?" Treeka interrupted.

"Don't worry. You won't have sex. He is looking for something else. Look, we don't have time for me to explain everything. Just follow my lead," Nozomi explained.

A feeling of shame arose as Treeka put on the provocative leather outfit, which showed all of her skin and barely covered her private areas.

Nozomi handed her an overcoat and a leather bag. "We will attract too much attention with these outfits, so you will wear this until we get to the target. The bag contains items you will need later," she said.

Treeka buttoned up the overcoat, which stopped just above her knees. Nozomi examined a red leather satchel, then strapped it to her waist. Then she put on her own overcoat just before locking the room.

"We have fifteen minutes to get to the alley—let's make it in ten."

Nozomi darted toward the exit before Treeka could respond. Treeka gave chase, and moments later, they were running together at a high rate of speed, opening doors and running through them faster than any human could. A map of the complex appeared on Treeka's HUD. She remembered that it had taken them over half an hour to reach the alley before, while walking at a brisk pace; this time, the duo made it there in four minutes.

A black van with four men in black ninja-looking outfits greeted them in the alley. It was still dark outside, but the men wore black sunglasses.

"Who are these guys?" Treeka asked.

"Nothing more than the hired help. Don't speak. Follow my lead."

"There she is!" a man with a British accent said. "Are you going to show us those leathers, like you did for the target in Belgrade?"

Nozomi smiled and opened her overcoat to show the man as he approached. He responded with a long whistling sound.

"It's good to be working with you again, Noz," the man said.

"Malcolm, I wasn't expecting to see you this evening," Nozomi said.

"I'm breaking in a new crew." Malcolm pointed to a man who was at least seven feet tall with a stone gaze. "That is Rocco Surrelli. He just arrived from the States to assist us."

The man nodded in response.

Malcom pointed to a much smaller man. "That small mousey-looking fellow is Teddy. He will provide your getaway car."

Teddy was much shorter than an average man; Treeka estimated that his height was less than five feet, since he was shorter than her. He wore a hunter's cap that covered most of his greasy hair. He lowered his glasses, winked at Nozomi, and then blew her a kiss.

"Don't be fooled by her good looks," Malcolm told Teddy. "Noz is as deadly as they come, and she's brought a friend."

Malcolm then pointed to a large brute of a man; he looked to be the most formidable in the group. "That balding bloke with the beard is Stump. He will accompany you to the target's room."

"Who is our target?" Treeka asked.

"Does she even know what's expected of her?" Teddy laughed to the other men. Then he looked back at Treeka. "You can practice here." He pointed to his crotch.

The other men laughed. Nozomi gave Treeka a look of

annoyance; Treeka decided it was best not to inquire any further.

I still don't know what I'm supposed to do—are we going to kill or torture this man?

"The target's name is Ioann," Malcolm said to the group. "He's a high-value target in a Russian crime syndicate known as the Collective, and he has critical information that Noz needs. He has ordered a lady of the night to pass the time. Nozomi and her friend will help Ioann in his time of need. He's into a lot of kinky stuff, so he will be expecting those leather outfits."

Stump opened the van door. As soon as Nozomi and Treeka were seated, the driver floored the vehicle.

The Thames was visible on Treeka's side of the van as they drove. She could see also Big Ben and the London Eye in the distance.

About thirty minutes later, Treeka noticed a familiar glow to the east.

The sun is rising.

"We will let you ladies off at the back entrance to the Mecco Hotel. From there, you will take a lift to the tenth floor. Ioann will wait for you at the end of the hall."

Twenty minutes later, Treeka and Nozomi knocked on Door 1004 in the Mecco Hotel. Stump waited for them by the elevator. At the end of the hallway, a window showed a picturesque view of London's east side. Treeka gazed upon the most beautiful sunrise she had ever seen. The pinkish clouds contrasted with the golden rays of the sunlight, bathing her in a wash of light. Nozomi knocked on the door again; no answer. Treeka gave Nozomi a sideways glance, and then the door opened.

A man who looked like he was in his late forties answered. His nose was red, and he kept rubbing it and making hooting noises. The man gave Nozomi a once-over. He seemed to be pleased to see her. He looked at Treeka and frowned.

"I only ordered one of you," he said.

"Consider it a two-for-one special," Nozomi said as she kissed the man with such intensity that he almost fell over. He stumbled back into the doorframe. Treeka followed.

"Wait—I can't take my reward until I have been punished," Ioann said.

Nozomi tossed the overcoat to the floor and then produced a red whip from her satchel.

The man held his hands together. "Bind me, mistress."

Nozomi gave Treeka a nod; she had told Treeka what she needed to do during the van ride. Treeka took a pair of leather handcuffs from her bag and bound the man. She attached a rope to metal loops that protruded from the handcuffs. Treeka examined the room. It was a loft with high ceilings.

The man looked around for a moment, then pointed at the ceiling. "Hang it from there."

Treeka threw the rope over a solid-looking rafter, then caught the dangling rope. Nozomi grabbed one end of the rope and bound the man's hands with it.

"Pull the rope," Nozomi demanded.

Treeka hesitated.

This is not me . . . What am I doing? Treeka thought.

"Remember, your sister is counting on you," Nozomi teased.

Ioann groaned as Treeka pulled the rope. His arms straightened as Treeka pulled. She could see his muscles straining to keep up with her pulls. Treeka pulled hard; the man screamed as it lifted him off the floor.

What am I doing? she asked herself again. Tears rolled

down her face, and her arms started shaking. *I must save Meeka
. . .*

"Help me tie it off," Treeka said.

Nozomi assisted Treeka in stabilizing the rope. The man writhed in agony as the weight of his body pulled his arms.

We will pull his arms from their sockets. He probably will enjoy it if that happens.

Treeka caught a glimpse of the man's face; he was wearing a mask of pain.

"Show me how much of a bad boy you've been," Nozomi teased.

"I've been a bad boy—I need more punishment," Ioann groaned.

Nozomi removed the sleeves from her leather outfit, revealing her milky, pale skin. She approached the man and kissed him on the lips. Then she bit him on the chest and neck —not once, but several times. He screamed; Treeka couldn't tell if it was from pleasure or pain.

"Pull harder," Nozomi said to Treeka.

Treeka loosened the rope, then pulled with all her might.

"Harder!" Nozomi said.

Treeka strained, but she managed to lift the man another half-foot off the ground. He screamed in delight.

Nozomi pushed Treeka out of the way and took control of the rope. She pulled so hard the man lifted another foot. His screams grew shriller.

He must be in agony.

Treeka wiped away more tears. She moved away from the man. She watched in horror as Nozomi hit, bit, and kicked the man repeatedly.

"I've been bad. I need mother's milk," Ioann said in a weak voice.

"Take this," Nozomi said to Treeka, who assumed the burden of the rope.

Nozomi removed her top, revealing two perfectly round breasts. The man licked his lips in anticipation. Nozomi took back the rope then pulled harder, her white hair covering part of her face. Nozomi's smile widened as the man wept. Treeka regained the rope and pulled the man a few more inches off the ground.

"No, my baby," Nozomi cooed. "Don't be sad. Mommy is here."

Nozomi dragged over a small chair and stepped upon it. She gave the man a hug. She ran her fingers through his hair, then stopped. She stepped and turned away from the dangling, screaming man.

"Face me," the man said in a low voice.

"I cannot hear you, maggot," Nozomi yelled.

"I desire to be with you, mistress," Ioann said.

Nozomi was silent. She let the man hang for several moments. He started whimpering.

What the hell are we doing with this man? Aren't we supposed to eliminate him . . .

Treeka gave Nozomi an urgent look. Nozomi only nodded in response.

"Are you ready to tell your mistress what she desires?" Nozomi said.

"Yes," Ioann said in a weak voice.

Nozomi gave Treeka a lowering gesture. Treeka reduced the strain on the rope a bit. Nozomi held a hand up. Treeka got her meaning. She secured the rope to one of many vertical structural beams in the loft.

"I want you, mistress," Ioann said.

Nozomi moved closer to Ioann and removed the lower part of her white leather outfit, revealing a perfect female body.

Treeka couldn't tell if she was a cyborg; she had no cutouts in her flesh like Treeka did. Treeka took in every inch of her.

"Hold me," Ioann begged.

Nozomi crept closer to the whimpering man, lingering just out of reach. He tried to get closer to her, but the rope prevented any forward movement. Nozomi reached up for the Ioann's bindings, pressing against him as she stretched. He started kissing her arm.

"Stop!" she snarled. "You haven't earned that yet."

He stopped. Nozomi stretched up so she could whisper into his ear; then he appeared to say something to Nozomi in a low voice.

Treeka tapped her left earlobe to activate the implant.

"It's part of the machine now," Ioann said.

"Which machine?"

"Colossal . . . Machine," the man choked.

"Let him down," Nozomi said.

Treeka undid the rope and let it go. The man plummeted to the floor. Nozomi jumped off the chair and crouched before him.

"Give me more," Nozomi whispered.

Treeka barely heard the command, even with her enhanced hearing. The man appeared to be in terrible shape. Every vein on his body stood out, and he was sweating everywhere. Nozomi kicked him in the face. Blood splattered against her perfect white skin. She kicked him again and again.

"Ready to talk now, worm?" Nozomi said as she pressed her foot against his face.

The man was quiet; he just whimpered. She lifted the man's head. Her breasts were in his line of sight.

"If you want some of this, you better give me something first, my dear," Nozomi said.

The man pointed to his mouth. She kneeled down to listen.

The man grabbed her hair and pulled. Nozomi hit the man in the chest with a single motion. He stiffened, then stopped moving.

"Is he dead?" Treeka asked.

Nozomi checked for a pulse, then shook her head.

"He's out . . . for a while. I will give him some time to rest, then we will try again."

Moments later, Nozomi reached into her bag, produced a syringe, and stabbed the man in his chest. He gasped.

"Mistress is not letting you rest, you swine," Nozomi told him.

The man rolled over onto his back.

"I'm ready to tell," he whimpered. "Then I get my reward?"

Nozomi straddled the man, then gave him another passionate kiss. He started whispering again.

"The kid from the States holds the key," he said.

"What key?"

"To the AI you seek."

Nozomi pulled down the man's trousers. The man gasped.

"What are you doing?" Treeka demanded.

Nozomi shot her a murderous glance.

Treeka backed off.

She must be enjoying this.

Treeka gazed out the window. She was treated to a gorgeous view of the city. Treeka didn't want to hear the sounds of pleasure and pain from the man any longer. She just wanted her sister back.

CHAPTER 4

Four hours later

Treeka glanced back at Nozomi and Ioann. Nozomi was dressing in silence. Ioann handed Nozomi something that resembled a flash drive. They embraced for a long moment.

"Let's go," Nozomi told Treeka.

Ioann licked his lips, then smiled. He took a seat on the couch. Treeka lingered a moment longer, staring at the remains of the rope.

What just happened? This all seems very . . . unreal.

She had to run to catch up with Nozomi, who was waiting at the elevator in the hallway with Stump.

"What the fuck was that about?" Treeka asked.

"I suggest you ladies keep your voices down," Stump said.

The elevator opened; a couple who appeared to be in their early twenties gave Stump a wary look, then exited. They didn't even look at Treeka or Nozomi. Stump inspected the elevator, and then the cyborgs entered. Treeka noticed that Stump stood just inside the elevator's doors, discouraging others from entering. He pressed a button marked "LL."

"Not all of our work is killing and mayhem. Sometimes you need to use your other gifts," Nozomi told Treeka.

"I'm not doing that kind of work when I'm in the field—damn the consequences."

"We all do things we don't enjoy, especially for those we love," Nozomi said.

Who does Nozomi love? Treeka wondered. *She seems all alone in this world.*

Treeka followed Stump and Nozomi to the hotel's service entrance. A familiar van was waiting for them. Rocco opened the door, and Malcom greeted the pair.

"Ladies, I trust your meeting was productive?"

"Very productive," Nozomi said.

"I bet—Ioann is going to remember this for a long time," Teddy said, laughing.

Malcolm and Rocco joined in the laughter. Whatever inside joke that amused the men was lost on Treeka.

"Did you get what you came for?" Malcolm asked.

"Yes—Mother will be so proud," Nozomi replied with a smile.

About forty-five minutes later, Nozomi and Treeka strode in the underground lair of Dr. Ash's lab.

"I don't understand what happened back there," Treeka said.

"The mission was to get information from Ioann, and that's what we did. He even gave us a bonus payment," Nozomi said as she showed her the flash drive.

"What's on it?" Treeka asked.

"Untraceable cryptocurrency called Digibit—payment for services rendered," Nozomi said, smiling.

"Was it necessary to do those things to him? I thought we were just supposed to eliminate him."

"Eliminating him wasn't necessary because he was cooperative. The primary objective was to gather the information. He was willing to pay for certain acts that I'm good at," Nozomi said.

"He could have given you bogus information. What if he tells his boss?"

"I know men, and his information is solid. I can always tell when men are lying. Especially men in vulnerable positions. He won't risk telling his boss."

"How does this help my sister?"

"It doesn't, but it allows us to secure funding that we need to keep this complex operational. And help your sister."

Twelve hours later

Treeka and Nozomi sat at an abandoned loading dock near Dr. Ash's lab. Treeka smelled the faint charred smell of burned plastic and other debris. The women had an unobstructed view of the London Eye; it was just across the Thames. She crossed her arms and shivered, and it wasn't from the chilly night air. She felt disconnected from her previous life. The more she tried to access her memories, the more they fluttered away like fall leaves in the wind.

Movement—someone was nearby. Treeka leaped to her feet and readied a dagger she had tucked in her belt. Nozomi put a hand over her blade and made a lowering motion with her other hand.

"He's here," Nozomi said.

"The courier?"

Nozomi nodded.

"Hello, ladies," a man said.

Treeka looked around; she couldn't see anyone. Moments later, a man of medium height approached from behind a shipping container. He was short, clean-shaven, and dressed in an overcoat.

"Jackson?" Nozomi said.

"It's me, Noz." Jackson produced a cylindrical object. "We need to get this to Dr. Ash."

"Hands where I can see them," Nozomi said.

Jackson put his hands up as Nozomi frisked him.

When Nozomi was satisfied, she led Jackson and Treeka through a loading dock that looked abandoned. She unlocked and opened a metal door leading into darkness.

Thirty minutes later, the image of Dr. Ash was waiting for the unlikely trio in the dimly lit lab. Overhead lights turned on as Nozomi, Treeka, and Jackson arrived.

"Hand it over," the image of Dr. Ash said.

A white robot on wheels liberated the canister from Jackson and took it into a nearby room.

"Once the sample stabilizes, we can begin our work on restoring your sister," the image of Dr. Ash told Treeka.

Treeka nodded. She hoped they could bring her back.

Jackson shed his overcoat and gloves and hung them from a nearby chair.

"So, what now, ladies?" Jackson said, rubbing his hands together. "Can you turn on the heat?"

"Not in the lab. It is at a constant temperature of 4.44 degrees Celsius to keep our delicate materials from getting spoiled. They will deteriorate at higher temperatures," Dr. Ash said.

Jackson nodded as he rubbed his arms to warm himself.

"Did you have any trouble getting the sample?" Dr. Ash said.

"I separated the genetic material before Norris got to the lab in Munich. There were casualties at the lab," Jackson replied.

"That is regrettable, but the important thing is that you have the sample," Dr. Ash said.

"Did the information from that agent pan out for you?"

"I'm afraid it raised more questions than it answered. Now we need to find some kid from the United States. Nozomi was able to acquire more names from the agent, which led us to a gaming company in Munich."

"Pretzelverse Games?" Jackson said.

"Yes, that sounds about right."

"The genetic material came from a lab belonging to a sister company of Pretzelverse. I know the company hires a lot of teenage interns from all over the world."

"Interesting. I will have Nozomi investigate that angle," Dr. Ash said. "But for now, you must be exhausted from your journey. Nozomi will show you to your quarters."

"Wait—I wasn't planning on staying here," Jackson said.

"I require your help, Jackson," Dr. Ash replied. "It's not like I can perform the procedure myself. My robot helpers can only do so much, and I don't want to damage Treeka's sister any more than is required."

Treeka glared at Jackson, who took a step back.

"Nozomi will take care of . . . all your needs. If you need anything, all you need to do is ask," Dr. Ash's image said as she floated toward some robotic workers.

"I will give you two days, and then I'm on a plane to Munich," Jackson said.

"Thanks for agreeing to help, Jackson," Dr. Ash said, before adding, "and I'm sure Treeka appreciates it, too."

Treeka gave Jackson a "don't-screw-this-up" look.

"Why don't you rest while I make my preparations? I need you fresh for tomorrow," Dr. Ash said.

"I'm yours in case you need help with your preparations, Jackson," Nozomi said as she caressed his face.

"Oh, that is kind of you to offer, but I think I have everything I need. Just . . . show me to my room if you don't mind," Jackson said.

"Follow me," Nozomi said, smiling.

Jackson followed Nozomi down a darkened hallway and disappeared into the night.

"What can I do to help?" Treeka said.

"Try to think of any stimuli we can use to help Meeka awaken from her slumber. She's in a comatose state, but she can still absorb certain sensory details like sound or touch. Anything like that is helpful in case we need it," Dr. Ash said.

"I will clear my mind and think on it," Treeka said as she headed toward her room.

The next morning, after Treeka checked her fluid and food cartridge levels, she strode into the lab feeling hopeful. Jackson, Nozomi, and Dr. Ash gathered around a table containing a naked and unconscious Meeka. Treeka noticed that metal was visible in several spots around her midsection. It looked like the skin was peeling away where the metal protruded. She touched her metal side unconsciously. Jackson was wearing a pair of AR glasses and operating some sort of portable tablet.

"Preparing for the disposition voidance procedure, doctor," he said.

"What's that?" Treeka asked.

"It's a little too complex to discuss with a layperson," Dr. Ash said.

"Try me—I want to understand what you are doing to Meeka," Treeka said.

The floating screen displaying Dr. Ash's image seemed to roll its eyes; the sight would have been comical under other circumstances.

"It's a procedure that will help Meeka's neural pathways and synapses join with her cybernetic frame. The procedure was developed by the real Dr. Ash and another scientist named Ron Allison. Meeka's brain is used to being in her human body. This procedure helps her brain accept her new cyborg body."

"We need to prepare the subject first. Her brain requires a jump start with the material, which asks like a catalyst," Jackson said.

Jackson's AR glasses chirped. He tapped them before moving out of earshot from Dr. Ash's crew.

Jackson better not be getting called away—what is he up to?

Treeka followed Jackson as he entered a corridor near the lab. She was eager to learn more of his conversation, so she activated her enhanced hearing implant.

Voices—they are too low. I need to adjust, Treeka thought as she tried adjusting the implant.

"The real Dr. Ash is dead? I'm with the replica now. She's performing some impromptu science experiment on a cyborg," Jackson said in disbelief.

Who is he speaking with?

"Got it. No—the replica of Dr. Ash does not suspect. I will start Project Longevity as soon as I hang up. I need to get back before I'm missed."

Treeka felt a presence: it was Nozomi, just behind her. The cyborgs exchanged a look. Treeka pointed in Jackson's direction. Nozomi used two fingers in a walking motion and pointed in Jackson's direction.

"Follow?" Treeka mouthed.

Nozomi nodded.

They followed Jackson back to his quarters. There were no windows, and Jackson wasn't speaking with anyone else, so Nozomi motioned Treeka to follow her. After several minutes of traversing empty hallways, Treeka followed Nozomi into a room with a battered metal door. Nozomi turned on the lights, and Treeka gasped as she took in the array of monitors and boxes with blinking lights in racks.

"From here we can watch Jackson," Nozomi said.

Moments later, the pair were watching Jackson's every move. He appeared to be in his quarters. Nozomi zoomed in on his laptop screen. Treeka could make out certain words such as "Longevity" and "transference." The more she watched, the less she understood what he was doing.

"What's Jackson doing? I don't understand," Treeka said.

"He's shutting Dr. Ash out of the mainframe," Nozomi replied.

"What's a mainframe?"

"It's a giant computer that our Dr. Ash is connected to. Without it, she cannot function. Your sister cannot come back."

Treeka clenched her fists and punched a nearby metal table that collapsed from her outburst. A group of monitors also crashed to the floor.

"Careful—we can't get your sister back if you damage our equipment," Nozomi scolded.

"We can't let him do this. What are we going to do?" Treeka said.

"We will call Dr. Ash—she's good at these things," Nozomi said while pulling up a communications panel.

Moments later, the image of Dr. Ash filled one monitor. Nozomi informed Dr. Ash of the situation.

The image of Dr. Ash smiled. "Perfect—now I can get my body back."

"How are you going to do that? Won't it rot before we can get to it?" Treeka asked.

"The real Dr. Ash has contingencies in place. She will be preserved, but time is still against us. We have less than two hundred hours before the body will complete its deterioration," Dr. Ash explained.

"What are we doing about Jackson?" Nozomi said.

"Don't worry—Jackson's communications have been intercepted. His mainframe access has been revoked; he won't know it until Meeka is awake. We still need his help for the procedure. We won't be able to bring Meeka back without him," Dr. Ash said.

"What do you want us to do?" Treeka asked.

"Nothing. While we've been talking, Jackson left his room. Both of you get back to the lab—now."

Jackson better not do anything else to prevent Meeka's return, or he will pay, Treeka thought.

Several hours later, when Meeka's vital signs had stabilized, Dr. Ash called everyone into the preparation room. Treeka's sister was resting peacefully on the table, surrounded by machines. Treeka examined the series of electrodes and wires that were connected to parts of her head. The wires led to several machines displaying graphs, numbers, and charts.

"Jackson, I will need your help in stimulating Meeka's

temporal lobe. I will introduce some light music into the chamber to help open her mind. The human brain, specifically the temporal lobe, responds positively to music. On my mark, I will need you to start the procedure," Dr. Ash said.

"I'm ready," Jackson said.

Nozomi and Treeka watched from behind a glass partition. A light melody played as Dr. Ash attempted to stimulate Meeka's brain. The song began with a light piano, and then more complex instruments like a bass guitar and drums were introduced. The melody wasn't classical or soft rock, but rather a combination of the two.

The music increased in intensity; the few instruments gave way to a full orchestra. Treeka thought she recognized the music from a dance class her mother had taken her to as a child.

Meeka doesn't like this music, Treeka thought. She likes punk and dance music.

"The subject is not responding to the stimuli, attempting level two," Dr. Ash said.

"Isn't that risky? We should try less evasive actions first," Jackson said.

"We need to introduce electroshocks to the hippocampus portion of the brain; this will work with the music to provide additional stimulation," Dr. Ash said.

"How long do we have?" Jackson said.

"The genetic material has already started breaking down. We have less than an hour before we'll need to start over."

Treeka barged into the room.

"I know what the problem is," she said. "You need to play some music that Meeka likes. Play something that young people would dance to in a club."

"Do you have anything specific in mind?" Jackson asked.

"Anything with a beat—Meeka loved to dance."

Jackson changed the music to an electronic, industrial-sounding beat. To Treeka, it sounded like someone was pounding a piece of metal with a hammer. The beat of the hammer became more urgent-sounding as the tempo increased. Other sound effects were layered in as the song progressed.

"I'm showing some increased brain wave activity now," Dr. Ash said.

Jackson changed the tune to something resembling a humming and a steady drumbeat. Strange electronic sounds accompanied the rhythm. Slowly, other instruments like guitars and trumpets were introduced. Treeka thought she recognized a chainsaw.

Treeka looked at the monitor just above Meeka's body that showed her brain waves. When they had started the procedure, the line had appeared to be straight, with no variations. As soon as the industrial sound started, more spikes were present.

Jackson switched to a popular dance track that Treeka recognized from Meeka's many clubbing excursions. It had an intense beat, and two men took turns singing. One voice was deep, and the other was shrill. The deeper voice started singing about destruction while the shrill voice cried about being stuck in a time zone. The longer this song played, the more brain wave activity was present.

Meeka's responding to the music!

"The subject's heart rate is increasing," Jackson said.

Meeka's arms and legs twitched as the lyrics changed to blowing up people, and something about assassinating a president or a king.

Treeka remembered how Meeka loved dancing with just about anyone in a club. Once she had danced with an older man who had kept up with her every move. Meeka didn't mind being close to the men who danced with her. By the end of the evening, Meeka was usually dancing with three or four men who appeared to be all over the age spectrum. As long as they kept up with her wild dance moves, they were welcome to dance with her.

Meeka's eyes opened.

"Her eyes are opening. Shall I start the procedure now?" Jackson said in an urgent tone.

"Meeka has entered into stage-three REM sleep. We can start the procedure when she enters stage four—and not a minute sooner," Dr. Ash explained.

"But . . . her eyes?" Jackson said.

Meeka sat up and looked at Jackson with lifeless eyes.

"She's moving, Dr. Ash."

"You need to restrain her, Jackson," said Dr. Ash.

He tried wrestling her back onto the table. As soon as he got her down, Meeka's head turned to face her sister. Treeka put a hand up. Meeka stretched her arm toward her sister. When Jackson attempted to strap her left arm to the table, Meeka struck with the speed of a viper. She punched Jackson in the face with her free hand. Jackson stumbled back.

"Argh, you crazy bitch!" he yelled.

Meeka undid the strap holding her arm, jumped on Jackson, and then started biting his neck. The electrodes and wires still attached to her pulled one of the monitors to the floor. It bounced before sputtering out.

Jackson screamed. "Get her off me!"

Treeka started to intervene, but Nozomi held out a hand.

"No, sister. Let her become who she is meant to be," Nozomi said.

Treeka looked over at Nozomi, whose smile widened.

She's enjoying this.

Meeka started chewing on Jackson's ear. She bit off part of it, then spit it out on the floor. Jackson's screams turned into shrill cries. It sounded more intense and urgent than anything Ioann had let out.

"Retrain...the umbilical is still intact. I'm starting the disposition voidance procedure," Dr. Ash said.

Treeka noticed a thick, hoselike object that was attached to the base of Meeka's neck. Most of the electrodes were still attached. She stopped attacking Jackson when the disposition voidance procedure started; then she dropped to the icy floor and started flailing. Jackson ran out of the room, covering his mangled ear as he went.

"Mouth guard—we need to put it back in. She might bite her tongue off," Nozomi said.

Treeka ran to Meeka, found the mouth guard on the floor near her sister's spasmodic body, and shoved it back into her mouth. Then she put her sister's head onto her lap and tried to soothe her as the procedure continued.

An hour later, Jackson emerged with a bandage covering his ear. Dr. Ash floated in front of him. Treeka stayed by her sister, who was resting on the table.

"I thought you would like to know that the operation was a success," Dr. Ash said.

Nozomi leaned against the doorframe, watching Jackson with a predatory stare.

Jackson sneered. "Good, because I'm out of here. I want to be long gone when that crazy bitch wakes up."

Treeka shot a hateful glance at Jackson.

"She's not crazy, nor a bitch. You should watch your tone."

Dr. Ash floated between them. "Nozomi will give you a little extra something for your efforts," she told Jackson.

Nozomi produced a small flash drive. "Extra payment in cryptocurrency for services rendered—plus a little extra for your accident."

Jackson reached for the drive; Nozomi pulled it back and laughed. "You can't have it that easily." She unzipped her tight leather outfit down to her navel, revealing perfect, bare flesh. "Do you want to play with me?"

Jackson grimaced. "I'm in no mood for games—my ear is throbbing, and besides, you are not my type."

Nozomi smiled as she zipped up.

"Very well, but it's your loss. Follow me, please."

Jackson followed Nozomi out of the lab.

"You should go with them," Dr. Ash said. "Don't worry—you will be here when you sister wakes. I need you to see how Nozomi deals with rejection."

Treeka nodded, then headed in the direction that Nozomi and Jackson had gone. Soon she was in one of the long hallways that connected the subterranean sections of Dr. Ash's compound. As Treeka turned onto a long hallway, she saw the pair just up ahead. Nozomi was sauntering like a mistress entering a whorehouse. Jackson still didn't appear to be interested.

When Treeka was about thirty feet away from the pair, Nozomi spun around, produced two daggers from her belt, and then plunged them into either side of Jackson's neck. As she removed the daggers, a fountain of blood shot into the air. Nozomi kicked Jackson. He staggered back into a wall. Nozomi threw a knife into Jackson's chest; then she threw her remaining dagger into Jackson's groin. Jackson screamed for a moment, but it sounded more like a drowning gurgle. Nozomi grabbed his head and snapped his neck. She kicked him several more times, then spat on his body.

Why did she kill him?

Treeka's knees buckled, and she fell to the floor in horror. Tears formed in her eyes.

"Why?" Treeka pleaded.

Nozomi shot her a wicked grin. "Hello, sister. I just finished putting down a rabid dog that was threatening to bite us."

'Sister'? I'm in a sisterhood of pain and anguish, with a psychotic taskmaster steering us into misery, Treeka thought.

Moments later, Nozomi tapped her AR glasses. Two helper robots zoomed past Treeka and cleaned up the mess that had been Jackson.

CHAPTER 5

TREEKA AND NOZOMI walked back to the lab in silence. Treeka had been starting to trust Nozomi, but that had all evaporated when she'd murdered Jackson.

Why did she kill that man?

Dr. Ash's image greeted them as they reentered the lab.

"Nozomi, I'm so proud of you. I appreciate that you could dispatch an enemy with such efficiency."

"I thought Jackson was a friend?" Treeka asked.

Dr. Ash's image looked sad for a moment. "It's regrettable that a friend would turn on us, in our time of need."

"I took care of the bad man, Mother. I just hope my sisters will have the fortitude and conviction to carry out these tasks when the need arises," Nozomi said.

Treeka gave Nozomi a sharp look.

"I will do my part, but I needn't be so brutal," Treeka said.

"My methods get results. Other assholes will think twice before messing with Dr. Ash."

"Dr. Ash is dead—this is her image," Treeka said.

Nozomi hit Treeka in the throat. "Shut your insolent mouth!" she shouted.

Treeka fell to her knees; she was having difficulty breathing.

"Stop this. Unless you are training, I forbid you to strike any of your sisters again," Dr. Ash said.

"She would never survive in the Society," Nozomi said.

"Treeka's not like you—she is special in her own way. She is not a killer . . . yet," Dr. Ash said.

Several moments later, Treeka recovered. Her throat burned. Nozomi had made the mistake of turning her back on Treeka, who repaid Nozomi in kind with a kick to the head. Nozomi didn't see it coming. Treeka fell back into her combat stance and motioned Nozomi to attack.

"That goes for you too, Treeka—I will not have chaos here, not when we are so close," Dr. Ash scolded.

"Only a coward attacks from behind. You are an unworthy sister, and a disgrace," Nozomi snarled.

The cyborgs stood down, but Treeka kept a wary eye on Nozomi.

"I need both of you to help Meeka—she is waking up now."

"Sister," Treeka said as she pushed a robotic helper aside and embraced Meeka.

Meeka looked confused.

"I'm sorry—do I know you?" Meeka said as she sat up and held her head. A robot helped her untangle the myriad of wires and sensors attached to parts of her body.

"She doesn't remember you. Give her time. Remember your memories were fragmented, and still are," Dr. Ash said.

Dr. Ash and Nozomi were behind her; in her excitement, she had not heard them enter. Treeka's vision blurred, and she wiped the tears away from her eyes.

"Another weakness we need to work on," Nozomi said as she aggressively wiped a tear from Treeka's face.

"Treeka is holding on to her humanity, dear," Dr. Ash said. "In time that will become less of a liability."

Nozomi looked disgusted.

"The Society trained you to become a living weapon, but you still have emotions. They are suppressed, but there," Dr. Ash told Nozomi.

"Is that why you were transferred into this crude construct?" Nozomi pointed at the floating image of Dr. Ash.

"This form was necessary because my human counterpart was still functioning. Thankfully, my human counterpart expired in a lab, so technicians there could safely preserve the body until I can merge with it."

"How did she die?" Treeka asked.

"From what I understand, it was Delta-51—an earlier model from the autumn series of cyborgs I created. The Dr. Ash who recently vacated my body was helping a billionaire transfer his ailing granddaughter into a Delta-51 construct when it turned on her. Unfortunately, Delta-51 carved out the eyes, but I can grow more," Dr. Ash said.

Treeka was silent in shock; Dr. Ash seemed so calm about everything.

"I need you girls to be ready for what is coming," Dr. Ash continued. "I will send you to New York for your first solo mission soon. Since Nozomi is the most experienced she will guide you, but I would prefer it if you and Meeka could complete the operation without Nozomi's help; that would go a long way into getting you inducted into the Society," Dr. Ash said.

"I keep hearing references to a Society. What are you talking about?" Treeka asked.

Nozomi looked at Treeka like she was daft. "The Silent Assassins Society," Nozomi said.

Treeka laughed. "Is that the official name? It seems a bit on the nose—don't you think?"

"There's nothing wrong with the name," Nozomi sneered.

"The original name was 'the Order,' but everyone calls it 'the Silent Assassins Society,' or just 'the Society' for short," Dr. Ash explained.

"Okay, so when is this mission?" Treeka asked.

"In nine days, I'm afraid."

"That's not enough time to prepare. We still need to get Meeka trained. She's not used to fighting," Treeka said.

"That's not true. Your father trained both of you in jujitsu if I'm not mistaken. From what I understand, you are both good at blades. That is why I picked you," Dr. Ash said.

"Picked me? We were in an accident . . . before this . . ." Treeka trailed off.

There is so much more that I need to find out about myself and my past.

"Oh yes, you were, my dear, but let's not get off the subject," Dr. Ash said. "You have seven days to train your sister before you travel to the States."

Treeka looked at her sister.

"I will train Meeka myself, and we will bring Nozomi only when needed," Treeka said.

"Are you sure that's wise, young one? I have so much to teach Meeka—perhaps she will take to my lessons better than my previous apprentice," Nozomi said.

Treeka glared at Nozomi.

"Very well, I will let you train your sister," Dr. Ash told Treeka. "But if you discover any anomalies between you and Meeka, I trust you will let me know?"

"Anomalies? What are you talking about?" Treeka asked.

"During my early experiments, I discovered an unusual telepathic bond between subjects who had shared a strong emotional relationship before the enhancements. This bond is even more intense among siblings."

Treeka looked confused.

"Are you saying that I can possibly communicate with my sister without speaking?"

"All I'm saying is to keep an open mind. If you notice anything unusual, I would like to know about it."

Treeka nodded, then left with Meeka following.

The next morning, Meeka was sitting with Treeka at a small table in the middle of the lab, examining a food cartridge. She clearly didn't know what to do with it.

"Are you hungry?" Treeka asked, making an eating gesture with her hands.

Meeka nodded.

"We don't eat with our mouths anymore," Treeka said, pointing to the metal side of Meeka's torso.

Treeka tried to lift that side of Meeka's shirt to show her when Meeka grabbed Treeka's hand.

"No undressing . . ." Meeka trailed off as her lips started quivering.

She's taking this harder than I thought. But Meeka was always worried about her figure. The metal plate is upsetting her, Treeka thought.

"It's okay. I have one too," Treeka said, pulling up her shirt and revealing a metal plate.

Meeka copied her action. She touched the metal plate that covered the left side of her body; Treeka remembered how she had felt when she'd first seen it. At first, she was alarmed, but she got used to it. Then she remembered seeing Nozomi's naked body and how it appeared.

Why is her body so perfect and ours aren't?

Treeka made a mental note to ask Dr. Ash about it the next time she saw her.

Treeka removed her shirt and ejected the food cartridge. "When the fluid level gets to this line, then it's time to replace the cartridge," Treeka explained.

Meeka replaced a cartridge by herself and seemed to perk up a bit.

Next, Treeka demonstrated the use of some of the sparring equipment in the training area. After a few attempts, Meeka was proficient with everything her sister taught. One time Meeka disarmed Treeka, who was holding a sword. Her fighting stances were perfect; it appeared that she remembered her childhood training after all.

I'm glad my sister is back, but I hope her memory returns soon.

By the end of the day, Treeka and Meeka were practicing their combat training. The girls didn't hold back; Meeka came at her sister with a katana. Treeka had two daggers and dodged or deflected each blow. The fighting increased in intensity, and one of Treeka's daggers got knocked out of her hand. Meeka resumed her fighting stance, then motioned for Treeka to retrieve her weapon. Once Treeka had her weapons, the fighting resumed.

Sparring with her sister brought back so many memories. Her father would often observe the twins' sparring sessions. He would offer pointers after the session was over, but he would never interrupt. Treeka missed those days, but she always came up short in meeting his high expectations. Her father held himself to the highest standard and expected nothing less from his daughters.

Still sparring with Meeka, Treeka suddenly froze, blinded by a mysterious flash of light; the training arena dissolved around her, and a new scene emerged before her eyes. She saw herself outside a bar, getting into a car with her sister and two other men.

Am I dreaming?

She was watching herself but couldn't interact. Everyone was wet from the rain. Meeka laughed and drank from a bottle. The driver took the bottle from Meeka, then drank as he drove. Moments later, Meeka passed out in the front seat. The man in the back seat grabbed the bottle from the driver. It had a pungent odor, but the man in the back seat kept drinking. The driver stopped at a red light; Treeka couldn't see anyone else at the intersection. The other man urged him to run the light. The car pulled forward, then stopped in the middle of an intersection. The rain intensified, and Treeka couldn't see anything. The men started arguing about something. Treeka saw lights— then darkness.

How long has it been since the accident? she wondered as *the vision dissipated.*

Treeka's memory was fuzzy around the time of the accident; it was almost as if that part of her memory had been wiped. Fragments of memories would make it through, but she couldn't put them together. She remembered getting hit by another car, but everything after that moment was blank—until the lab.

"Are you okay?" Meeka asked, noticing Treeka was lost in thought.

"Yes," Treeka replied. "But let's take a break."

Treeka bowed to Meeka. She returned the gesture. For the briefest of moments, Treeka thought she saw something in her sister's eyes: familiarity.

Is she starting to remember now too?

Another memory of a younger Meeka in a schoolgirl

uniform bowing to her appeared in Treeka's mind's eye, and then it too was gone.

After a brief rest and a break to check food and fluid levels, Treeka tossed Meeka a bow. Meeka gave her a confused look. Treeka showed her how to use it, and Meeka nodded with recognition.

The bow was one of Meeka's favorites—I hope that jogs her memory.

Meeka hit each target with precision. Treeka blindfolded Meeka, then helped her position the bow. She took her time, and Treeka noticed that she was controlling her breathing before each shot. She didn't miss a single target. She screamed in delight when she took off her blindfold. The sisters embraced for a long moment.

She will remember who I am—she just needs more time.

After several days of training, Treeka thought she saw Nozomi sneaking glances at their training sessions. She would hide behind a dummy or punching bag, but Treeka still noticed her.

Soon Meeka seemed to remember how to use the sword, bow, and small arms like pistols.

I can't believe she is picking this up so fast, Treeka thought.

It's like riding a bike, sister, Meeka's voice replied in her head.

What? Treeka froze. Was that Meeka?

Treeka gave her sister an appraising look; she just smiled back.

Something is going on.

"Can you show me around the facility?" Meeka asked.

"Sure—we can use a fluid break, anyway," Treeka replied.

Treeka took her sister to the kitchen—or what passed for

one. The area resembled a standard kitchen, and it contained a sink, a refrigerator, and another device that was almost as large as a standard oven. Treeka opened a small door on the oven-shaped object. Several cartridges were visible. She checked their food cartridge levels before replacing them. She poured two gigantic glasses of water and handed one to Meeka. Both emptied the glasses, and Treeka put them in the dishwasher.

"How long have you been here?" Meeka asked.

Treeka's brow furrowed. "I'm not sure, but it's been a while I think."

Meeka's eyes narrowed in concentration, as though she was starting to retrieve a memory.

As they left the kitchen, a small robot ran into Treeka.

"Ouch!" she cried. "Watch where you're going, you stupid butler robot."

Treeka rubbed her shin, then heard the most delightful sound. Meeka was laughing.

"Thanks for rubbing it in—I'm in serious pain here," Treeka said before joining her sister in laughter.

Tears came to Treeka's face as she laughed with her sister. She peered into Meeka's eyes; she was experiencing the same emotions.

"I'm remembering now," Meeka said suddenly. "Who is Huruto?"

Meeka became silent as she strained to recall more. Treeka didn't interrupt; she wanted her sister to experience the emotions as she recalled her thoughts.

A wave of regret washed over Treeka at the mention of Huruto. Most of her memories had flooded in soon after Dr. Ash had activated her. That name belonged to her father. She remembered having powerful feelings for him when she was reactivated, but after each passing day, the warm thoughts and feelings were becoming increasingly rare.

"Huruto is our father," Treeka said.

Meeka didn't show any emotion.

I know that Meeka and Father didn't get along, but I know she loved him. Are her feelings gone? Treeka wondered.

"Are you ready to show me around now?" Meeka asked.

"Yes, follow me," Treeka urged.

Treeka gave her sister the grand tour. Meeka asked questions when she would see something interesting, like the robot lab. Treeka escorted Meeka through the labyrinth of hallways and alcoves. Then she passed the battered metal door Nozomi had led her through when they had spied on Jackson.

It might be worth checking in on Nozomi . . .

She attempted to open the door, but it wouldn't budge.

It is locked? How was Nozomi able to open it?

Treeka tried yanking the door with more force, but it still didn't move.

"Help me," she told Meeka.

Meeka helped, and the door opened. A screeching noise echoed through the hallways and the door strained.

After a moment, and when there was enough space, the sisters slipped into the door's open gap.

Treeka fumbled in the dark before she found the light switch. The monitors were off, and Treeka did not understand how to activate the surveillance system. She wasn't a techie, and Nozomi had operated it before. Meeka started examining the desk, then crawled underneath it. Moments later, all the monitors came to life.

"How did you turn it on?" Treeka asked.

"I just plugged them into the power outlet," Meeka said, smiling.

Treeka smiled back at her sister. She was glad to be reunited.

"This room monitors the entire facility," Treeka explained. "Nozomi showed me earlier."

"I don't like her," Meeka said.

The feeling is mutual, sister.

"There must be a dozen monitors here," Meeka said.

"I count thirteen, including the small one on the desk," Treeka said.

Each of the larger monitors had four squares showing different camera feeds.

Treeka examined the small monitor in more detail. She remembered Nozomi fiddling around with it earlier. As soon as she touched it, a grid of six squares appeared. She touched one of the six boxes on the smaller screen. All the feeds on the larger monitors changed. She repeated the process until she saw what she was looking for: she saw Nozomi and Dr. Ash in one square. They seemed to be in a heated discussion. It looked like they were in the lab.

"Can we hear what they are saying?" Meeka said.

Treeka looked for a volume dial or button that would let them eavesdrop on the conversation but couldn't find anything. Then she noticed a speaker-shaped icon on the small control screen. As soon as she tapped it, voices blared through the small room. After some quick adjustments with the volume control, they could hear the conversation.

"Why are we going to New York?" Nozomi asked.

"Because Nico Gratzano has taken over his father's empire. He is now in charge of the largest mob family in the United States. We need him on our side if we hope to get my human body back," Dr. Ash said.

"Wait—I thought your lab techs have your body," Nozomi said.

"They did, but as of this morning, it was missing from the facility where it died."

"How?"

"Norris, the man with the mustache. The one you lost from the hotel bar . . . He acquired it from the facility. Norris and his men slaughtered all of my lab technicians."

"I'm confused. How do you know all this?" Nozomi asked.

"Here," Dr. Ash said, pointing to a screen.

Dr. Ash pulled up a video on one of her monitors showing a massacre unfolding in a lab; several people wearing lab coats were getting mowed down by automatic rifles. Norris, the man Treeka had seen at the hotel bar with Jackson, was leading a group of commandos into battle. The images appeared to be coming from various body cameras and kept switching perspectives from each commando's point of view. It looked like some military operation.

"The final video gives us clues about the location of my body," Dr. Ash continued.

The last video was Norris explaining that he had secured Dr. Ash's body and would meet Jackson in Munich.

"If your body is in Munich, why are we going to New York?" Nozomi asked.

"Because the Gratzano family was hired to acquire it for a Mr. Tage. I found an email communication about Project Longevity on Jackson's laptop. There were several communications with a Mr. Tage. I was able to correlate his involvement to the Gratzanos and the meeting spot in New York."

"I will not even ask how you came across that information," Nozomi said.

"It wasn't hard once I regained control of the AI known as Leviathan from Delta-51. Now, the last piece of the puzzle is getting my body back within the next one hundred and one hours," Dr. Ash said.

"What happens if we don't?"

"My body dies—for good, this time. The lab technicians

preserved my body using a special cryogenic process that keeps the body fresh for almost ten days without a full cryo-unit. Five days have passed, so you have little time."

"What are we telling the twins? I mean, why aren't you telling them now?"

"While this is an important mission, it is also a test of their abilities and loyalties. The successful completion of this mission is not only vital for me; it is also crucial for the Society."

"When are we leaving?"

"Soon. I need you to watch the twins until you do," Dr. Ash said.

Are you ready for your first mission? Treeka thought, looking at Meeka. To her surprise and delight, Meeka's voice replied wordlessly in her head.

I don't trust Nozomi or Dr. Ash, Meeka said.

The twins exchanged a look that reminded Treeka of a secret the sisters shared as children: Meeka would raise an eyebrow so high that it looked like it might fly off her face. Whenever Meeka gave her that look, it meant that they should be on high alert. Meeka gave her that expression now. Treeka nodded in response.

Treeka and Meeka hurried back to the training area.

About an hour later, Meeka found the robot who controlled the music; it was one of those butler-looking robots. Meeka persuaded the robot into playing the familiar beats of dance and techno music. The music blasted through the training area, and Meeka started her dance moves.

She hasn't forgotten how to dance, Treeka noted.

During their limited down time, the girls would dance. Treeka didn't like dancing, but she remembered how much

Meeka had enjoyed it. Meeka took to the rhythm of the beat. As the music intensified, Meeka seemed to remember her old life.

"Come on, Treeka, dance with your little sister," she said.

Treeka started to dance, a bit more clumsily than Meeka's smooth dance moves. Meeka closed her eyes and let the music take her like she had so many times before.

Treeka sensed someone approaching from behind.

"We will not dance the enemy to death," Nozomi chided.

The twins turned and glared at Nozomi. She was watching them with some interest.

"No, but dancing is a good way to get close to someone. You should try it sometime," Treeka said.

Nozomi tried some moves, but they looked awkward to Treeka.

"No, that won't do," Meeka said.

Treeka just watched as Meeka gave Nozomi some basic lessons on how to dance. Nozomi seemed to have two left feet because she kept stumbling.

She's never danced with anyone before, Treeka figured.

After several attempts, Nozomi seemed to start to get the feel for dancing. The trio danced together for some time.

Nozomi looks like she might be having a good time—is that possible?

"Bravo," Dr. Ash said while playing some sound effects that resembled applause.

The three cyborgs gave Dr. Ash a look.

"How long have you been there?" Nozomi asked.

"Long enough to see you girls have some fun. It's about time," Dr. Ash said.

Meeka looks as annoyed as I feel, Treeka noted. Dr. Ash shouldn't have interrupted our bonding time.

No, she shouldn't have, Meeka replied.

Treeka looked in Meeka's direction, then gave her a conspirator's smile.

"You will travel to New York," Dr. Ash said.

Treeka did her best to feign surprise. "Why are we going?"

Dr. Ash's expression changed to something Treeka couldn't place. She didn't know if Dr. Ash had any human feelings, or if it was some elaborate programming. Treeka was curious if Dr. Ash could lie to them.

"I would have liked to have you girls train together for a long period, but something has come up and we must take swift action. Nozomi will fill you in once she puts together a tactical plan," Dr. Ash said.

Nozomi nodded.

"Change into your traveling clothes and meet back here in thirty minutes. We have already placed suitable attire in your rooms," Dr. Ash said.

The twins, Nozomi, and Dr. Ash regathered in the training area. The outfits that Dr. Ash provided looked like something that college freshmen would wear.

"I have arranged passage to New York City." Dr. Ash said. "The three of you will travel together to the States on a student visa. All the paperwork is ready."

"No one will believe we are students," Treeka said.

"All three of you are of college age. Nozomi is a little older, but she can pass for a graduate student. Besides, with your cybernetic conditioning, you won't age anyway, so that doesn't matter. The important thing is that you look the part," Dr. Ash continued.

"Not a problem, Dr. Ash, we got this," Nozomi said.

Meeka shot Treeka a glance.

Do you think we can pull this off, Treeka? Meeka thought.

So we can actually communicate telepathically, huh? Treeka asked.

I know—it's the strangest thing, huh? Meeka replied.

We've done it, Meeka. I didn't expect it to happen, but it did. We've established our telepathic connection. We can hear each other's thoughts.

Groovy. I wish we had that ability when those nasty men tried . . . Meeka lost her train of thought.

Treeka gave her sister their secret look.

Meeka, I don't think we should tell Dr. Ash—yet! Treeka thought.

Agreed. I don't trust anyone who doesn't have a body, Meeka replied.

"Yes," Treeka said, chuckling.

Dr. Ash and Nozomi gave her a curious look.

Didn't mean to blurt that out, Treeka thought.

"Yes, we got this," Treeka said.

That seemed to placate Nozomi and Dr. Ash.

"Do you remember what you need to do once you get to Heathrow?" the image of Dr. Ash asked.

Nozomi nodded. "We go through security and then meet Mr. Lowe, who has additional instructions for us," she said.

"Who's Mr. Lowe?" Treeka asked.

"Your handler. He makes it possible for us to travel to other countries. He provides cover stories and identities. Now stop interrupting," Nozomi said.

"Yes, it will be an excellent opportunity to introduce the twins to Mr. Lowe. They still have much to learn but will make a fine addition to our ranks," Dr. Ash said.

"Where do these kill orders come from? Is it possible to complete the mission without killing anyone?" Treeka asked.

Nozomi sneered. "You ask a lot of questions for someone who isn't a member of the Society. Know your place, little one."

"I'm not little," Treeka said furiously as she closed the space between her and Nozomi. "You have not made me feel welcome," she continued. "You have been undermining me since the beginning." Treeka was now standing mere inches from Nozomi, invading her personal space.

"Back off," Nozomi said.

"Girls, this behavior is not becoming of you. You cannot act like this in the field where one wrong move can get you killed," Dr. Ash said.

"We will do our part," Treeka said.

"I don't need any amateurs ruining the mission," Nozomi said as she left the room.

"Oh, we have some upgrades for you two," Dr. Ash said.

"Upgrades?" Meeka asked.

"Follow me—it won't take very long."

"What are you going to do?" Treeka said.

"I'm giving you partial access to my database," Dr. Ash explained. "While you don't have real-time access, I've included several relevant case files and other information such as what to order in the restaurant."

"Is Nozomi getting this upgrade?" Treeka asked.

"No—she already has a real-time uplink to my AI. The only thing she needs is Wi-Fi access, and she can establish a secure connection. You're getting a very limited subset. You can think of it as a portable encyclopedia."

The twins followed Dr. Ash to her lab. The procedure was about as invasive as getting a shot at the doctor's office. With the help of a robot, Dr. Ash installed something into the metal embedded into their skin. Treeka stared at it for a long moment.

"Don't worry. I'm getting you girls re-skinned when you get

back," Dr. Ash told her. "Then you will look as perfect as Nozomi."

It's about time we got some upgrades, Treeka thought. *But Meeka and I are not performing any sex acts for people like Ioann.*

We can have some fun on our own terms. Right sister? Meeka replied.

The twins shared a smile.

The next morning at Heathrow a tall, well-dressed man opened the door of the limo that had transported the cyborgs. He was middle-aged with a trimmed beard and, despite the cloudy sky, wore sunglasses.

"Welcome, ladies," the man said.

He took Nozomi's hand.

"I'm Mr. Lowe, your travel adviser."

Treeka and Meeka followed Nozomi out of the vehicle. The driver already had their luggage at the curb. Mr. Lowe paid the man, and the limo sped away. He handed each of the women an envelope.

"Your passport, with your identification papers, some cash, and plane tickets are in there. Be sure to use the last security checkpoint to the right side of the terminal. A security officer with a yellow tie will get you through security," Mr. Lowe said.

The trio thanked him. After checking in, filling out the appropriate visa paperwork, and getting their passports scanned, they found themselves in the security line. A rather large woman with a yellow tie checked their identification.

"Traveling to the States, I see. How long you staying?" the security officer asked.

"Just a few days. We are attending a conference," Nozomi said.

"Well, I hope you get to take in some sights."

Nozomi smiled, then after the three of them were checked in, the security officer directed them to a special security area. They were patted down, and their luggage was searched. Minutes later, the women were sitting in a bar close to their gate.

"Can we even drink alcohol?" Meeka asked.

"You can drink it, but it affects you more than a regular non-modified human. It also has a dehydration effect, which can affect your fluid levels. It is best not to drink, but if you must, have your drinks shaken: that will lessen the effects. A martini is a good drink for shaking," Nozomi said.

Meeka looked disappointed.

"We will be careful," Treeka said.

"Flight 48 nonstop to New York is now boarding. Please present your ticket and passport to an attendant no later than ten minutes before departure," an announcer said.

The flight was seven hours, which was both an exciting and exhausting experience for the cyborgs. They did not get tired like humans did, but they did require nourishment twice a day. Treeka passed the time watching old kung fu movies and playing games that allowed her to shoot fat birds into the air. It took an extra ninety minutes to get off the plane and clear customs once the flight landed.

An enormous bald man with no neck greeted Nozomi, Treeka, and Meeka just outside the customs area at JFK International.

"Rocco Surrelli, I wasn't expecting you to pick us up,"

Nozomi said. "It seems like it was just yesterday when you helped us with that agent in London. When did you get back?"

"A few days ago," Rocco replied, giving Meeka an appraising look. "I've heard a lot about you from Dr. Ash. It's good to finally meet you."

"It's good to meet your acquaintance," Meeka said.

Rocco smiled. "Well, let's get you ladies to the city."

He loaded their luggage onto a nearby cart, then tapped his watch. By the time they entered the outside pickup area, their limo was waiting. Another man who resembled a linebacker opened the back door for them.

"This is Sims," Rocco said. "He will take you to your place in the city. It's nice, and it faces Central Park: the perfect spot to relax and plan a takeover of one of the most powerful crime bosses in the tristate region."

Sims gave a nod as the women entered the car. Rocco loaded the luggage, said something to Sims, and then waved at the cyborgs. Moments later they were driving to the city.

It took over an hour to get into the city. Sims was mostly calm, but sometimes he took alternative routes to avoid the crowded parkways. Sims pulled up to a valet area outside their hotel, and an attendant parked their car. Sims carried in their luggage. As the cyborg trio entered the apartment lobby, Treeka noticed that they were attracting some attention from several onlookers —mostly from men.

Sims didn't allow anyone else to enter the elevator they used to get to the hotel's penthouse level. Once there, he escorted them to the apartment that would serve as their base of operations. He unlocked the room and then inspected it before letting the women enter.

"You will find everything you need in your suites," Sims said, and he was gone before the women could respond. Treeka looked out the apartment and noticed that Sims was entering the stairwell. The apartment was enormous; there was a full living room, bar, and several bedrooms. The cyborgs spread out and unpacked their luggage. Nozomi then started setting up a portable communications station. She unpacked three laptops and lined them up on a nearby desk. Meeka went to one of the bedrooms; moments later, music could be heard.

Treeka looked out of the penthouse's window. The sun was setting over the Manhattan skyline. Treeka had been to the city once before when her father opened his New York office. They had stayed in a five-star hotel, and this apartment was a lot more luxurious.

Treeka, do you think we can have some fun while we're here? I want to go to a club, Meeka thought.

We can ask, but I don't think Nozomi will approve, Treeka replied.

"What's our timeline?" Treeka asked Nozomi.

Meeka turned off the music and reentered the living room.

"We need to complete the mission and be back in London in sixty hours, hopefully before. So if you're thinking of doing any clubbing, you'd better put those thoughts out of your mind," Nozomi said while looking at Meeka.

Nozomi opened an enormous closet. Instead of clothes, weapons of all kinds were on several shelves. Treeka saw knives, swords, guns, and electric stun guns.

"In two hours, a rival crime family will dine at a popular karaoke bar. We need to take out Louis Diamond—"

"Who is Louis Diamond?" Treeka asked.

"Stop interrupting!" Nozomi paused for a moment before continuing. "He's the head of a minor crime family who is trying to control part of New York. There is another person yet to be identified. I will coordinate with the mother ship, and I will need to stay here while you ladies go in for the kill."

"Who is the other person?" Meeka asked.

"You deaf? I just said I don't know yet."

Treeka gave Nozomi a hateful look.

"I expect to have that information soon, so wear these so we can communicate with each other," Nozomi said, handing the twins earpieces.

"Any suggestions on how we should get close to Mr. Diamond?" Treeka said.

"He loves a pretty face, so it shouldn't be difficult to get close to him," Nozomi explained.

"When are we going in?" Treeka asked.

"Soon—I will let you know when," Nozomi said as she inspected her computers and surveillance equipment.

Meeka gave her sister a hesitant glance.

I got you, sister, Treeka thought.

Meeka nodded.

Later that evening, Treeka and Meeka entered the living room area of the penthouse. The twins were wearing black leather outfits that covered most of their skin and exposed metal. Their shoulders weren't covered, however, and their skin there was visible. After checking fluid and nutrition levels, the twins settled on the couch. Moments later, Nozomi entered wearing a silver evening gown. She looked like she was going out to a formal dinner. She had braided her long white hair, and it hung to one side.

"I approve of the look. You look like sisters of pain in that black leather," Nozomi said.

"Sisters of what?" Treeka asked bitterly. She shot Nozomi a

look of disgust; she hated remembering their encounter with Ioann. She didn't even know why she had been invited. She had learned nothing—except that Nozomi was a sadist.

"Never mind," Nozomi sighed. "You just look good, that's all."

Meeka pulled a sword from the closet. She practiced her stances. She looked like a natural; she seemed to be one with the sword.

"Too bad you can't go into the bar with that sword," Nozomi said.

"What weapons are we supposed to use?" Meeka asked.

Nozomi threw a small pouch to Meeka. She opened it and saw two small daggers.

"Attach it to your belt; these blades are small and easy to conceal," Nozomi said as she tossed another small pouch to Treeka.

"Besides Mr. Diamond, who else are we targeting?" Treeka asked.

"Rocco has that information. He's been doing a little reconnaissance work for us this evening. He should be here soon."

The twins attached the blades to their belts as instructed. Meeka practiced drawing the weapon in various positions. Treeka joined her.

Moments later, three knocks reverberated throughout the penthouse. Nozomi opened the door to let Rocco in.

"Noz, you look fine tonight," Rocco said. "But you are only a few years older than my son, so I better keep my distance."

Rocco laughed, but Nozomi didn't look amused.

"You got news for us?" Nozomi asked.

"I do. We are to keep Mr. Diamond alive long enough for him to tell us where Nico Gratzano is. He is the real target. We just need to go through his henchmen first," Rocco said.

"Are we targeting anyone else? We only know about Mr. Diamond."

"Oh, let me see." Rocco rummaged through his coat pocket and produced a note. "The second guy is Mikey Lombarto, Louis Diamond's cousin. We can kill him anytime. He doesn't know Gratzano."

Rocco pulled out his smartphone and showed Treeka and Meeka photos of their targets.

"I will monitor police bands and restaurant feeds from here," Nozomi said while opening a laptop.

Rocco's smartwatch chirped. After a quick glance, he motioned for the twins to follow him.

"Diamond has just been seated. We should go, ladies."

"Remember, I can monitor your communications. I will send reinforcements if you get into trouble," Nozomi said.

Treeka and Meeka followed Rocco out of the penthouse and into the night.

CHAPTER 6

Rocco drove south in his black sedan toward the twins' first challenge. Meeka was staring out the window; she seemed envious of all the shoppers on Fifth Avenue.

"You girls ready?" Rocco asked.

"We will do our part," Treeka replied.

"It's okay to be nervous, being that it's your first time. But I just wanted to give you a word of advice. Keep your heads down and do as you're told: the Society will reward you. Cross the Society at your peril," Rocco said.

"Are you threatening us?" Treeka said.

"No, I'm just trying to help. I was in another group called the Collective who would stop at nothing to get what they wanted. They were a brutal group with no honor. The Silent Assassins Society is the real deal. They treat their people good, pay well, and if you do the same, they will look after you. That's all I'm saying."

"Thank you, Mr. Surrelli," Meeka said.

"Just call me Rocco—Mr. Surrelli was my father."

"Once we dispatch Mr. Diamond, what is our exit strategy?" Treeka asked.

"Oh, we've got that covered. How are your motorcycle

skills?"

Meeka gave Treeka a smile.

Remember our races? Meeka asked.

Yeah, Jonny Boy didn't think we could handle those crotch rockets, especially on Mulholland Drive, Treeka replied.

"Yes, we know how to ride them," Treeka said. "Why?"

"Good—we have two bikes stashed two blocks north of the bar; it's better that you use those," Rocco replied. "I've sent the rendezvous point to your smart glasses. Confirm that you received the instructions."

Treeka removed her AR glasses from her purse and put them on. She tapped them, and a basic interface appeared. It showed a map with two pins: one labeled "bike," and another labeled "meetup."

"I see a map with two points," she said.

"Good," Rocco said. "Your glasses can bring up any information that Nozomi sends you. You already have the neural cybernetic implants; just think about what you need and your glasses will show you. It's not as good as having a built-in AI, but it's better than nothing."

Rocco dropped the twins just down the street from the bar. It was snowing lightly.

"We're here. You girls will do great. I can feel it," Rocco said just before driving into the night.

Treeka and Meeka strode into Matzie's Karaoke Bar off Thirty-Second Street in Manhattan. The snow melted off their glistening black leather outfits; the pair looked identical.

An Asian woman, around thirty years of age, greeted them. "Would you ladies like to be seated?"

Treeka and Meeka gave each other a look, then Treeka nodded. The hostess seated them in the karaoke section of the bar. Meeka put a hand on Treeka's shoulder and pointed at the back of the bar. A few patrons were in that section, but a private party was occupying a room nearby.

I have a feeling that Louis Diamond is in that private party, Treeka thought.

Agreed, Meeka replied.

"We would like to be seated there," Treeka said, pointing in the general area of the party.

The woman looked confused. Treeka and Meeka shared a knowing glance and then began walking in the party's direction. The hostess blocked the entrance to the room and said, "This is a private function. I'm sorry, but it's off-limits."

"Acknowledge your statement," Meeka said.

The woman gave the twins a worried glance that suggested she was dealing with two escaped mental patients.

"Follow me, please," the hostess said.

Treeka and Meeka did as instructed. The hostess seated them at the edge of the bar; Treeka sat at the first chair that was available. She still had a line of sight into the other room. This would have to do for now.

Several minutes later, an older waitress introduced herself as Linda and asked if they wanted to order drinks.

"Bourbon—neat," Treeka said.

"And you, miss?" the waitress asked.

Meeka looked at Treeka for a moment. Her left eye twitched for the briefest of moments, and then she said, "Same—bourbon, neat."

Linda left with a puzzled look and mumbled something under her breath. Treeka rolled back the auditory footage that Dr. Ash's enhancements had provided and played back their encounter with the waitress for their collective brains.

"She muttered: 'Crazy fucking foreign tourists!'" Treeka said quietly.

Meeka looked annoyed; Treeka clarified with silent defi-
nitions.

*Crazy: A person not in their right mind. Fucking (human
slang): A vulgar swear word. Tourist: Someone who is not local
and visits for nonbusiness pleasure reasons.*

"So let's have a pleasant time tonight," Treeka said.

"Yes, sister," Meeka replied.

Treeka nodded and said silently, *Our telepathic bond is
getting stronger.*

Meeka smiled, then silently replied, *Indeed, sister—but I'm
not as advanced as you.*

Thirty minutes later, the twins were sipping their drinks. They
declined food when it was offered.

"We're just here for drinks," Treeka told the waitress.

Linda shrugged, then left the two of them to enjoy their
drinks.

The party in the other room was breaking up. Treeka
scanned the room next to the party room. Although they didn't
have a direct line of sight, they used the reflections in the
napkin dispensers on the bar to act as spyglasses. The images
were distorted, but Treeka could use her image-enhancing
receptors to compensate.

"Targets acquired, recommend full scan for confirmation,"
Nozomi said through their earpieces.

Then the people in the room began to disperse and spill out
into the bar. A large Italian-looking man emerged from the
room.

A message appeared on the twins' AR visor.

Target confirmed: Louis Diamond—target assigned to Treeka.

The twins waited until the room emptied more. Meeka's target was still not in sight. A few minutes later, more guests left the party room. Louis Diamond was still chatting with his friends, but he looked like he was getting ready to depart. About a minute later, Louis headed for the door with his conversation companion behind him. Treeka stood up and approached Louis with her bourbon in hand. She did her best to feign spilling the drink on Louis.

"Mister, I'm so sorry," Treeka said.

Louis looked into Treeka's eyes.

"It's okay, no harm done, babe."

His companion laughed. Then the companion slapped Treeka's behind. "Oh—a nice firm ass," he said. "My name is Mikey—nice to meet you."

Treeka gave the man a lifeless, icy stare. He retracted. "I'm sorry—I should have kept my hands to myself."

Treeka heard a loud gong sound that seemed to come from everywhere. An AR system message appeared on the twins' visors.

New target acquired: Mikey Lombarto.

Meeka was at Treeka's side in an instant. The twins shared a determined glance.

Not yet, sister, Treeka told her.

Treeka leaned over to Louis Diamond and whispered in his ear, "Do you want to take me home?"

Louis smiled. "Yes—I would like to see where this goes," he said.

"Why don't you tell me where Nico Gratzano is?" Treeka continued.

"Why would I?" Louis asked.

"Oh, Louie, I want a threesome with you and Nico, that's all," Treeka said.

"I don't think that's a good idea. Why don't you come home with me first, and then I might call Nico later, if I feel like it?"

This guy isn't falling for my usual charms—I'm switching modes, Treeka thought.

I will be ready, Meeka replied.

"Give me Nico, and I may spare your life," Treeka said.

"Fuck you, bitch," Louis said.

The twins both took out a pair of daggers from either side of their large belt buckles. Treeka plunged both of her daggers into Louis's neck. Blood splattered her face and bare shoulders, shooting out like a busted fire hydrant.

"Father!" a girl in her early teens yelled.

Meeka thrusted one of her daggers at Mikey, but he saw it coming and dodged. His sudden movement caused him to trip, and he stumbled into one of the restaurant's windows. Although Treeka was blocking Meeka's path to Mikey, she calculated the best path around her sister. The move was a blur to Mikey. Meeka appeared with both daggers in hand and a murderous grin on her face. She plunged both daggers into Mikey's chest, and the motion propelled him through the glass. Meeka jumped through the broken window and onto Mikey, then stabbed him until he wasn't moving. Blood pooled down the front of Mikey's white shirt.

The twins were aware of the gasps and screams emitting from the bar.

You did well, sister, Treeka thought.

Thank you, sister, I had a good teacher, Meeka replied.

Moments later, the twins heard sirens in the distance. A message on their HUD read:

Authorities immanent, do not get apprehended at any cost.

Treeka heard the whimper of a young girl who was kneeling beside Louis Diamond's corpse. She appeared to be in her early teens.

"Do you know where Gratzano is?" Treeka asked while pointing her bloody knife at her.

The girl held out a smartphone that looked like it was drenched with blood. Treeka snatched it from the girl's hand. It was unlocked, and she flipped through the contacts until she saw several contacts with the last name of Gratzano. Her new enhancements allowed her to send the information to Nozomi. She received a data upload confirmation, then threw the phone back in the girl's direction.

The twins ran two blocks down the street, activated their hidden motorcycles, then left the area.

About forty-five minutes later, the twins met Rocco at the rendezvous point in Central Park, near a baseball field. The twins hid the motorcycles between two dumpsters. Treeka found a stained shirt in one of the dumpsters and did her best to wipe the remaining blood off their outfits and weapons.

As the twins got closer, they could see Rocco near the pitcher's mound. He was throwing a ball into the air and doing his best to catch it. Treeka noticed that he missed most of the time.

"Girls, so glad you could make it. Nozomi has confirmed from the police scanners you did your job. Did you get the information that we require?" Rocco asked.

Treeka pointed to her head. "It's in here."

"Good, I'm glad it went well."

"What now?" Treeka asked.

"Nozomi is planning our next move. You need to give me what you scanned," Rocco said, handing her a pad of paper and a pen.

Treeka wrote out the Gratzano information and handed it to Rocco. He took a picture with his cell phone.

"I sent this to her already," Treeka said.

"Just confirming, standby."

Seconds later, his phone chirped. He texted something back. Treeka noticed his expression changed to a look of worried resignation.

"What is it?" she asked.

"Nothing to be concerned about—Nozomi is tracking down Gratzano now. Come, I will take you back to the penthouse."

The twins followed Rocco back to the sedan. He opened the door, motioning for the girls to enter. As Meeka made her way to the open door, Rocco pulled a pistol out from his pants and shot Meeka point-blank in the face. Treeka watched in horror as her sister fell and hit the ground. Meeka gave Treeka one last look before expiring. Treeka thought she heard something turn off in her head; it was like someone had turned off a radio in her brain.

I will kill that fucker if that is the last thing I do.

Treeka flung one of her daggers at Rocco; it grazed his neck. He ducked behind the open car door, just out of Treeka's grasp. Tears blurred Treeka's vision, but she pursued Rocco.

He must die!

She saw Rocco twist something on his wrist, and before she knew it, Rocco was suddenly behind her. Moving quickly, she turned to plunge a knife into his heart when she felt the air leave her lungs as he punched her chest. It felt like someone had hit her with a sledgehammer. She fell to her knees. Rocco punched her in the face. Her vision faded into nothingness.

Treeka opened her eyes. She couldn't move her head, arms, or anything else. Her only view was the ceiling.

I'm paralyzed.

She heard voices; they sounded familiar.

"We caught the people responsible for the deaths of your family leaders. As a gesture of goodwill, we are prepared to hand over the one that is still alive," a woman's voice said.

I know that voice—Nozomi?

"Oh yeah," a male voice said, "what do you want?"

"We need something from Nico Gratzano—something he stole from us."

"What is it?"

"A dead body of an older woman. It's my grandmother, and I want a proper burial for her."

Murmurs of conversation followed. Treeka couldn't make out any of what was being said.

"I can convince Nicky to let your grandma go, but I need to inspect the package first," said the male voice.

"All yours, boys!" Nozomi said.

The face of a man appeared. His head was long, as though someone had stretched it. He had a pointy nose with a scar running across it and black spiky hair.

"Ahh, this one looks tasty," the man said.

The man moved closer. He was a couple of inches from Treeka's face. He took in a long breath, as if he was trying to breathe her in. Then he licked the side of her face. She moved her head slightly—just enough to bite the man on the cheek. She wanted to chew his face off, but he pulled away.

"Argh, the bitch is feisty too. I like her. Consider it done—I will call you in the morning with the transfer details," another man's voice said.

"Can we do it tonight?" Nozomi asked.

"Don't think so—Nico just buried his father, so we need to be delicate on our approach, but I think we can do this tomorrow."

"The sooner, the better, as the service is in less than forty-eight hours. The family wants an open casket," Nozomi said.

"I understand. A proper burial for loved ones is important. I will be in touch soon."

Treeka heard a door open and close, and then Nozomi came into view. She gave Treeka a sadistic smile.

"Thank you for your participation," Nozomi said. "Dr. Ash will be pleased that your mission was a success. She won't be happy to lose two assets, but let's be honest, you're not much of an asset."

"Why?" Treeka asked.

"The whole 'hunting of Nico Gratzano' business was getting a bit tiresome, and we were running out of time, so I did the next best thing. I traded his rival's killers for Dr. Ash's body."

"I'm going to kill you," Treeka said.

Nozomi laughed, then produced a large needle and poked it into Treeka's neck.

"Good night," Nozomi cooed. "Have fun getting tortured, raped, or any other torment that Vince has in store for you. Oh, I forgot to tell you—he is Louis Diamond's brother."

Treeka heard the sounds of laughter as Nozomi faded away.

I hope you enjoyed reading Synapse of Ash. I invite you to continue Treeka's adventure with **Echoes of Silence**. The exciting follow-up to **Synapse of Ash**.

I invite you to join my reader group to learn more about this upcoming book. To sign up visit:
https://cyberhunterorigins.com

Order Echoes of Silence and it will be delivered to your eReader.

ACKNOWLEDGMENTS

Developmental & Copy Editing by Hayley Evans
Proofreading by Sandra Ogle

Cover Design by Konstantine Designs
    Paperback Cover Design by Warren Design
    Special Thanks to my advanced reader and launch team.

D. B. Goodin has had a passion for writing since grade school. After publishing several nonfiction books, Mr. Goodin ventured into the craft of fiction to teach Cybersecurity concepts in a less-intimidating fashion. Mr. Goodin works as a Principal Cybersecurity Analyst for a major software company based in Silicon Valley and holds a Master's in Digital Forensic Science from Champlain College.

Cyber Teen Project
White Hat Black Heart
War With Black Iris
The Making of Cyber Teen Project
Reckoning of Delta Prime (Summer 2021)

Cyber Overture
Sonorous
Chromatic
Resonance
Ensemble
Ramble

Interstellar Online
Blast Off
Cassidy's Fleet (Fall 2021)

Cyber Hunter Origins
Synapse of Ash
Echoes of Silence

Catalyst of Pain (Fall 2021)
Silent Assassins Society (Summer 2022)